Everyday Chaos

Published by arrangement with
UniPress Books Ltd, London, by the MIT Press

For image copyright information, see p.256

Publisher: Nigel Browning
Commissioning editor: Kate Shanahan
Project editors: Clare Saunders
and Richard Webb

Designed by Koen Slothouber and Sandra Zellmer

ISBN 978-0-262-53969-2

Library of Congress Control Number:
2020932919

Printed and bound in China

The MIT Press
Massachusetts Institute of Technology
Cambridge, Massachusetts 02142
http://mit.press.mit.edu

Everyday Chaos

The Mathematics of Unpredictability, from the Weather to the Stock Market

Brian Clegg

Contents

Stock Market Crashes and Super Hits

Harnessing Chaos

Complexity and Emergence

Introduction

Scientists and mathematicians have a tendency to take words that are used loosely in the world at large and give them very specific meanings. So, for example, in everyday speech we tend to use "power" and "energy" nearly synonymously. But in physics, power is very specifically the rate of transfer of energy from one location to another. Similarly, the words "chaos" and "complexity," which will be at the heart of this book, are broad descriptive terms in general usage, but in math they have meanings that imply particular characteristics.

In normal usage, when we speak of chaos, we think of a mess. A lack of order. Randomness. That word came to the English language via Latin, taken from the name of what some considered the primal Greek god, who represented the first, formless matter. The original word in Greek could also mean a chasm—but in either case, it denoted a lack of structure. Chaos spread confusion and was a force for destruction. It wasn't on the side of the good guys, which makes it an interesting choice as the term used to describe the mathematics of a surprising number of everyday things around us. First applied to animal population growth and the weather, chaos in the mathematical sense is typified by a system—a collection of things that interact—where very small changes in the way things start out can have huge implications for the way things eventually unfold.

If chaos implies unpredictability—disorder arising from apparently ordered starting points—the mathematical concept of complexity is a kind of *alter ego* (even though chaotic systems can be complex). In a complex system, the interaction of apparently simple components results in outcomes that would not otherwise be possible. Complexity is the ultimate end of "The whole is greater than the sum of the parts."

In ordinary usage, "complexity" simply refers to being made up of large numbers of parts or having an intricate form. But mathematical complexity can emerge from a relatively small

→

Untangling of Chaos
Sixteenth century engraver Hendrick Goltzius's illustration for Ovid's first-century epic poem *Metamorphoses*.

OVIDII METAM. LIB. I.
1.
E tenebris deforme Chaos secessit oborta
Luce, suoq; loco sunt queq; Elementa locata
Astra polo radiant, quibus imminet igneus Æther,
Aera subsequitur Pontus, subit vltima Tellus.
F. Estius

system, just as much as it can from an intricate mechanism. So, to be mathematically complex, a system doesn't have to be, well, complex.

One mark of a complex system is emergence. This is where the "greater than the sum of the parts" bits comes together. Emergence suggests that new abilities *emerge* spontaneously from the complex system without any guiding force being totally responsible for shaping those abilities. You, for example, are a complex system. If we take a look at the individual atoms that make up your body in its entirety, they are not alive. But you are alive. If we step it up a level, we could describe the cells in your body as alive—but they certainly aren't capable of thinking or feeling or carrying out the actions that your body does. These capabilities are emergent from the complexity that is you.

Perhaps what is most remarkable about chaos and complexity is that they are all around us in nature. They are present in every living thing, in the weather, in the majority of the real-world objects that we interact with. And they are there in many human creations and systems, from the stock exchange to a bookstore. Yet we don't get taught about chaos and complexity at school. They don't feature in a whole lot of the work carried out by scientists either, who often concentrate instead on the small details, producing results that aren't applicable holistically.

Much of science can be described as reductionism—breaking a complicated thing down to its components and seeing how those components work, then building back up from the individual parts to try to understand the whole. For example, a real-world chemical reaction can be chaotic. Anyone who has ever added concentrated sulfuric acid to water will know that the result is highly dependent on how you start out. But when we study chemistry, we break things down to their component atoms and only consider how they interact.

The twin theories of chaos and complexity give us the opportunity to get a better understanding of the real world, rather than the toy universe in which most science takes place. The real world is far more complex, chaotic, and, frankly, interesting than much of the science we were taught in school suggests. We are about to dive under the surface and discover reality.

Welcome to everyday chaos.

Que será, será

For the past two and a half thousand years we have developed an increasingly scientific viewpoint, often supported by mathematics. In some cases, this approach has proved remarkably effective. However, all too often the real world has confounded the attempts of science to predict what will happen.

It was not until the second half of the twentieth century that we realized what was occurring. The interaction of the components of systems, from an apparently simple jointed pendulum to immensely detailed weather systems, produces unexpected results. At the same time, collections of simple entities are capable of remarkable feats—think, for example, of the abilities of some species of ant that as individuals are totally incapable of any useful action but, by working together, can use their bodies to form bridges, stitch leaves, and carry weighty loads.

To see how chaos and complexity came to be understood, we first have to take a journey back in time to a point where it seemed that the future was entirely within the grasp of our mathematical minds. Thanks to the work of Isaac Newton, his successors were convinced that it would soon be possible to take on the universe and win.

Murmuration of starlings
Complex, shifting shapes emerge from the interaction of the birds in flight.

1
Clockwork and Chaos

Newton, Laplace, and the amazing clockwork universe

"An intellect which at any given moment knew all of the forces that animate nature and the mutual positions of the beings that compose it, if this intellect were vast enough to submit the data to analysis, could condense into a single formula the movement of the greatest bodies of the universe and that of the lightest atom; for such an intellect nothing could be uncertain and the future just like the past would be present before its eyes."
Pierre-Simon, Marquis de Laplace, 1749–1827

The fluidity of time
In our time, when technology is such a normal and everyday part of life, it can be easy to forget what a transformative piece of technology the mechanical clock was. Prior to the availability of clocks, time was a thing of approximations, with only the broadest reference points. Available to everyone was the apparent transit of the Sun through the sky, or the motion of the heavens at night (unless it was cloudy). The better-off might have had a sundial, a water clock that measured time through a liquid dripping out of a small orifice, or the progress of a candle's burn. But any sense of exactness with respect to time was not a real thing. This is apparent from certain enduring expressions, such as "the sands of time," referring to an hourglass, or the practice of fixing time by the Sun when using the terms dawn, midday, and dusk.

Our lives are now so tied to technology that precision in time can seem a burden—a thing of deadlines and pressures—but when mechanical clocks were first invented, they were instead a wonderful eye-opener. It wasn't just the ability to know what time it was—to be able to meet someone at a particular time, rather than having to wait an hour or two—it was an essential both for the daily observations of religions, which tended to be tied to specific times, and for science. Having a measured approach to the progress of time was crucial to beginning to understand how aspects of the universe worked. It is no coincidence that the great breakthroughs in grasping the physics of motion came about in Europe at the same time as relatively accurate mechanical clocks were becoming more widespread.

The earliest mechanical clocks seem to have been developed in Europe in the fourteenth century. It's difficult to pin down a first, but certainly one of the oldest examples was the tower clock of St. Alban's Abbey in England, constructed by Richard of Wallingford in the 1320s. This example did not survive the Reformation, but another early English clock, in Salisbury Cathedral and dating back to around 1386, is still in action. Like many clocks of the age it had no dial—the point of having the clock was for it to strike a bell on the hour to ensure that religious services, which were scheduled at specific hours of the day, could be performed on time.

The escapement—the mechanism that measures out the units of time—in these early clocks was inaccurate by modern standards. Relatively precise measurements of time were not possible until 1656 when Dutch scientist Christiaan Huygens invented a clock with a regular beat provided by a pendulum. A contemporary of Sir Isaac Newton, Huygens was among those who were driving physics in a more mathematical direction.

A few decades earlier, when Galileo Galilei had needed to time objects in motion, he had had to rely on imprecise measures such as his own pulse. But with Newton, mathematics took on a central role in explaining the universe—requiring the kind of accuracy that Huygens's clock and its successors could provide. Clockwork not only provided the beat against which motion would be measured, it gave a mental model on which to base understanding of the universe itself.

Mr. Newton's legacy

Since the ancient Greeks first studied the night sky the universe had been seen as resembling a mechanical structure, but with crystal spheres carrying the planets and stars, rather than the gears of a machine. We know the Greeks constructed a geared model (presumably not the only one) that reflected some of the motions of the heavens in the remarkable Antikythera mechanism, an astronomical calculator from *ca.* 100 BCE, discovered in a shipwreck off the Greek coast in 1901.

More dramatic astronomical clocks, of which one magnificent example is the *Orloj* in Prague, in the Czech Republic, dating from 1410, presented a clockwork analog of the universe, while small-scale devices known as orreries provided heliocentric models of the universe itself (what we would now call the solar system), showing the positions and orbits of the planets and moons, usually driven by a clockwork mechanism.

→

Prague Astrological Clock (Orloj), detail
Dial showing the paths of the Sun and Moon, the phases of the Moon, and more.

→ →

Prague Astrological Clock (Orloj)
Laplace's mathematical view of the universe is likely to have been influenced by the mechanical precision of astronomical clocks, such as this magnificent example from 1410 on the Old Town Hall in Prague.

PHILOSOPHIÆ

NATURALIS

PRINCIPIA

MATHEMATICA.

Autore *J S. NEWTON*, *Trin. Coll. Cantab. Soc.* Matheſeos Profeſſore *Lucaſiano*, & Societatis Regalis Sodali.

IMPRIMATUR.

S. PEPYS, *Reg. Soc.* PRÆSES.

Julii 5. 1686.

LONDINI,

Juſſu *Societatis Regiæ* ac Typis *Joſephi Streater*. Proſtat apud plures Bibliopolas. *Anno* MDCLXXXVII.

←

Newton's *Principia*
Title page of a first edition of Isaac Newton's masterpiece *Philosophiae Naturalis Principia Mathematica*, published in 1687 (the book should have been published in 1686, but the budget wasn't available).

These were "physical" models in the commonplace usage of that adjective. But with the work of Isaac Newton a new kind of model of the universe became available to natural philosophers (the name by which early scientists were known)—a *mathematical* model. Newton was not the first to describe the physics of motion—Galileo, for example, had made a start by studying the way that balls accelerated down ramps under the influence of gravity. However, Newton turned what had been primarily a descriptive science into one where mathematics could be used to determine the future.

In his masterpiece, *Philosophiae Naturalis Principia Mathematica* (Mathematical Principles of Natural Philosophy), usually known as the *Principia*, Newton used mathematical tools to describe how the attractive force of gravity between two bodies—for example, Earth and the Moon—caused them to move in a particular orbit and caused objects like the famous apple to fall to Earth. He also proposed his three laws of motion, explaining the way that bodies move and how forces cause them to accelerate and interact.

To achieve this feat, Newton developed a new type of mathematics, which he called the "method of fluxions," now better known as calculus, the name used by his competitor, the German polymath Gottfried Leibniz. Equipped with Newton's new mathematical tools, his successors were ready to take on the whole universe—and none more so than his most enthusiastic European supporter, the French natural philosopher Pierre-Simon Laplace.

No need for that hypothesis

Newton, in his work on gravity, had focused on the movements of bodies in the solar system. Laplace had a grander vision. Born into an aristocratic family in 1749, he showed an early talent for mathematics, which would blossom as he took on many of the problems of applying mathematics to the universe and bringing math to practical uses in physics and engineering. From our viewpoint, though, Laplace's greatest contribution was establishing the concept of determinism.

Here, the mathematical description of reality is taken to the extreme. Laplace envisioned the image of a clockwork universe, where everything that happens for all time is determined exactly by what occurred the moment before, proceeding mechanically under Newton's laws. To illustrate the implications of such a vision, Laplace dreamed up his "demon." His description in 1814 of this creation, which he described as "an intellect," is quoted at the beginning of this chapter.

According to Laplace's view, if someone knew every detail of the state of the universe at a point in time, then thanks to the mathematical certainties of Newton's and Laplace's mathematics, we would be able to predict perfectly what would happen next from moment to moment. Everything, for eternity, would be preordained.

As human beings, we have always wanted to know the future. Ancient civilizations had their oracles and auguries—mystical and magical means to get an apparent glimpse of what was to come. Astrology, until late medieval times, was considered an acceptable part of the scientific armory, on the assumption that the movements of the planets had an influence over what happened on Earth; it was frequently consulted by kings and commoners alike. By building on Galileo's observations, Newton was able to push aside the mysticism and make mathematical predictions that were of a different order to those of the oracles and astrologers. They worked. Repeatedly, *repeatably*, they foretold what would happen.

Newton's math described not only how things around us moved, but it tied the familiar movements of things on Earth to the apparently grander and totally separate movements of the heavens. He showed how the journey of the Moon around Earth, for example, could be predicted from the simple factors of the masses of the two bodies and their distance apart. Others would take this even farther. Newton's friend and supporter Edmond Halley (it was Halley who ensured that *Principia* was published) used Newton's mathematics to make an accurate prediction of the return of the comet now known as Halley's comet. He would not live to see its triumphant reappearance, but Halley's forecast was sound. The comet came back—on time.

Laplace went farther still. It seemed to him that, given perfect knowledge of how things in the universe were at a particular point in time, for every future time it should be possible to run the mathematical model of the universe forward and see what would happen next, moment by moment. It was a picture of a clockwork universe that ran on unwavering tracks. Yet for many it seemed implausible. How could reality be so far from this mathematical ideal?

→

Astrological zodiac
Fourteenth-century Spanish illustration of the signs of the zodiac, describing when the Sun enters each sign.

Taur
Leon
Scorpion

Randomness is predictable

"In that case, I would rather be a cobbler, or even an employee in a gaming house, than a physicist."
Albert Einstein, 1879–1955

→
Albert Einstein
Despite being a founder of quantum physics, Einstein (pictured here in 1921) was reluctant to accept the randomness at the heart of the theory.

Randomness in reality
For Laplace's vision to work, everything needed to follow on from what came before, with clear cause and effect, from moment to moment. It was a universe we would now describe as deterministic, meaning that everything that happens now is determined, clearly and unequivocally, by everything that happened the moment before. However, there was a clear problem blocking Laplace's view of the universe—randomness.

The idea that things can happen at random with no prior reason is not one that comes naturally to human beings. We understand the world around us through patterns, finding it difficult to accept that things can happen without a guiding principle—with no reason *why*. This dependence on patterns is excellent for survival when it comes to recognizing a predator or a dangerous situation. But it also means that we see bogeymen when there is nothing there, or assume, for example, that a disaster has to be blamed on the direct action of deity, or fate, or the malevolent intervention of a magical power.

In reality, from the days of antiquity it was realized that some events were, to all intents and purposes, random—the toss of a coin or roll of dice, for example—which is why such events feature in games of chance. But our disinclination to accept

randomness in the real world also explains why we tend to be taken in by processes based on such randomness, whether it's in gambling halls or the expectation that after a particular "run of luck" we can expect a change of fortune.

As mathematics became more powerful, it became apparent that, although randomness was by definition unpredictable in terms of the specific outcome of a single event, it behaved predictably over the long run. So, for example, we can't say what number will come up when we throw a (fair) die. It could be anything from 1 to 6. But we do know that each of the numbers should come up around one-sixth of the time. So, after a large number of throws, we would expect each of the faces of the die to have been thrown approximately the same number of times. We can't predict a specific outcome, but we can predict long-term behavior, and the more an event is repeated, the more accurate that prediction becomes.

Quantum confusion

The importance of randomness to science was brought to the fore in the early part of the twentieth century by the development of quantum theory. The physics of *very* small things, such as atoms, electrons, and photons of light, quantum theory shows that much of the behavior of these particles is dependent purely on probability. Each event is totally and inherently unpredictable in its outcome—but, over time, the pattern of outcomes is entirely predictable.

Photons
Although it had long been assumed that light was a wave, quantum theory made it clear that light also acted as a stream of particles, called photons.

This randomness caused Albert Einstein, one of the founders of quantum theory, to be so doubtful about it; he spent decades looking for flaws in the theory that he would never manage to uncover. Take an apparently simple example—reflection off a glass window. When light hits a piece of glass, most of it goes straight through, but some of it reflects back. This happens all the time but is particularly obvious when one side of the glass is more strongly lit than the other—for example, when you are inside a lit room at night. Trying to look out of the window, instead you see a clear reflection of the room you are in, not the world outside.

Although you see the room reflected back to you, this is not because light from the room is no longer passing through the glass—most of it still is. What you are seeing is the small proportion that is reflected back. (This happens all the time, not just at night, but during the day it is washed out by the strength of light entering from outside.) Probability comes into play when we consider the individual photons, the particles that make up the light that is streaming out of the room. When a particular photon reaches the glass, there is something like a 95 percent chance it will pass straight through and a 5 percent chance it will reflect back. But we can never predict what a specific photon will do.

Einstein felt that there had to be "hidden variables"—information that was available to the photon but not to us that allowed it to "know" what to do when it reached the glass. This is why he made a number of remarks casting doubt on quantum theory, including his comment about being a cobbler quoted on page 22, made in a letter to fellow physicist Max Born. But there is now unquestionable evidence that this is not the case. The photon has a probability of reflecting and a probability of passing through—but until one or the other occurs, nothing decides the outcome. It is truly random.

The power of probability

"Chance favors only the prepared mind."
Louis Pasteur, 1822–1895

What are the chances?
Games that rely on probability for their outcome date back to ancient times. It's not recorded who first spotted that a two-sided coin could be spun in the air and produce a reasonable approximation to a random selection between heads and tails, but coin tosses have certainly been used for random selection, fortune-telling, and games of chance for thousands of years.

Similarly, dice—or their early equivalents—go back a long way. Archaeologists have found astragali, shaped knucklebones that act as crude dice, dating back at least 5,000 years, while backgammon-like "tables" are among the oldest-known board games. For much of the time such methods of chance have existed, good players may have had an instinctive feel for how probability worked, but the rules of probability only began to be quantified when a sixteenth-century Italian physician, Girolamo Cardano, himself an enthusiastic gambler, wrote a book on the subject.

→
The Backgammon Players
Painting from 1634 by Flemish artist Theodoor Rombouts showing a lively game of backgammon.

Although Cardano was in his twenties when he wrote *Liber de Ludo Aleae* (Book on Games of Chance) and continued to refine it through his life, it was not published until 1663, nearly a century after it was first penned, because its topic was not deemed suitable for polite society.

Probability is a concept we seem naturally to struggle with, and Cardano's great contribution to help us understand it was to turn probabilities into more manageable fractions. Let's imagine that you had a fair coin and tossed it 100 times. Heads or tails? You would expect to get heads around 50 times and tails around 50 times. Cardano realized that you could represent the chance of getting a head as $1/2$ and the chance of getting a tail as $1/2$. The bigger the value, the more likely the outcome, with a value of 1 meaning absolute certainty and 0 something that would never happen. As a coin toss (ignoring landing on the edge) has to result in either a head or a tail, the chance of getting either a head or tail is $1/2 + 1/2 = 1$.

Similarly, with a familiar six-sided die, the chance of getting any particular number is $1/6$. Once our gambling scholar took a mathematical approach to probability, he could start to look at how to combine different outcomes to find the chance of a given combination. So, for example, if you want to know the chance of getting either a five or a six, it's simply a matter of adding the probabilities, giving us $1/6 + 1/6 = 1/3$.

Cardano also dealt with the distinctly trickier problem of combining multiple probabilities. To take a common example, we know the chance of getting a six with a single die is $1/6$. What's the chance of getting a double six with two dice? Cardano showed that this was a simple multiplication problem: $1/6 \times 1/6 = 1/36$. But what about the chance of getting at least one six with a pair of dice? Clearly it's a better chance than getting a six with just one die. But we can't double the $1/6$ chance—otherwise, rolling six dice would guarantee getting a six, which we know isn't the case. Cardano engaged in some lateral thinking. The chance of not getting a six with the first die is $5/6$. The chance of *not* getting a six with the second die is also $5/6$. So the chance of not getting a six with both dice is $5/6 \times 5/6 = 25/36$. As the total of all possible outcomes has to be 1, this means the chance of getting a six is $11/36$.

The power of distributions

Turning our minds to randomness and probability, the outcome is often less straightforward than it is in games of chance where the probabilities of the different outcomes are known in advance (assuming that the coin isn't double-sided or the dice loaded).

Equipped with the ability to master probability using Cardano's mathematical approach, there's an essential requirement for getting a handle on randomness and probability in the real world: an understanding of distributions, that is, a picture of how possible outcomes are distributed—whether, for example, some outcomes are more likely than others.

If we look at the distribution of possibilities for a coin toss, what we have, effectively, is a bar chart with two possible outcomes. We can approximate to the bar chart by tossing a coin repeatedly and noting down how many heads and tails we toss. Initially there may be significantly more of one than the other, but over time, as tosses are collected, the numbers will get ever closer to the expected distribution.

Similarly, we can produce an equally dull distribution for the possible results of throwing a six-sided die—and again, we could build this distribution without knowing the actual values by repeatedly throwing a die. Things get a little more interesting if we look at the distribution of values we can produce by rolling two dice simultaneously and adding the results together. The outcomes range from 2 to 12—but not all possibilities have an equal chance of turning up. Here the distribution is not only more interesting, it can tell us something—for example, that the most likely outcome of rolling a pair of dice is 7.

In the world around us, where something is varying randomly, we often find that the distribution comes in the form of a "normal distribution," sometimes called a bell curve because of its shape when plotted. So, for example, if you plot the height of a whole lot of people, you will find that heights of men and women are each distributed in an approximate normal distribution.

Normal distribution
Also known as a Gaussian distribution, the normal distribution is symmetrical on either side of the most likely outcome, with long shallow "tails" to either side showing outcomes that are less likely.

Any particular individual will, of course, only have one specific height—but what the distribution enables us to do is to predict the most likely height and how likely it is that a man or woman's height will be within a certain range either side of the most likely height. The normal distribution has a measure known

Distribution of coin tosses in an experiment after 100 tosses
As more and more coin tosses are made, the distribution will get closer to 50:50 heads and tails.

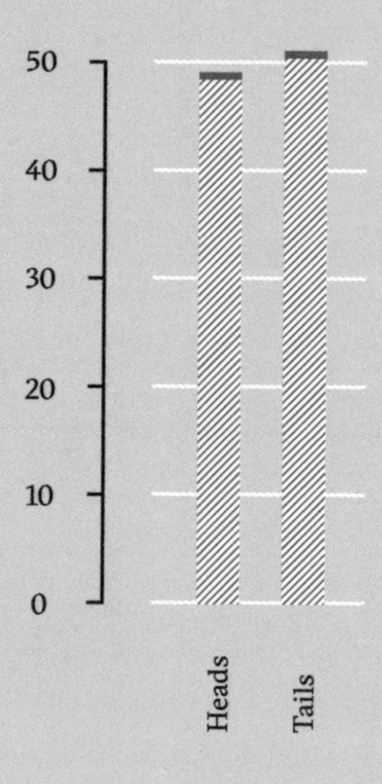

Distribution of the number of ways an outcome can be achieved with two dice.
Not all outcomes of throwing two dice are equally likely: the distribution shows relative probabilities.

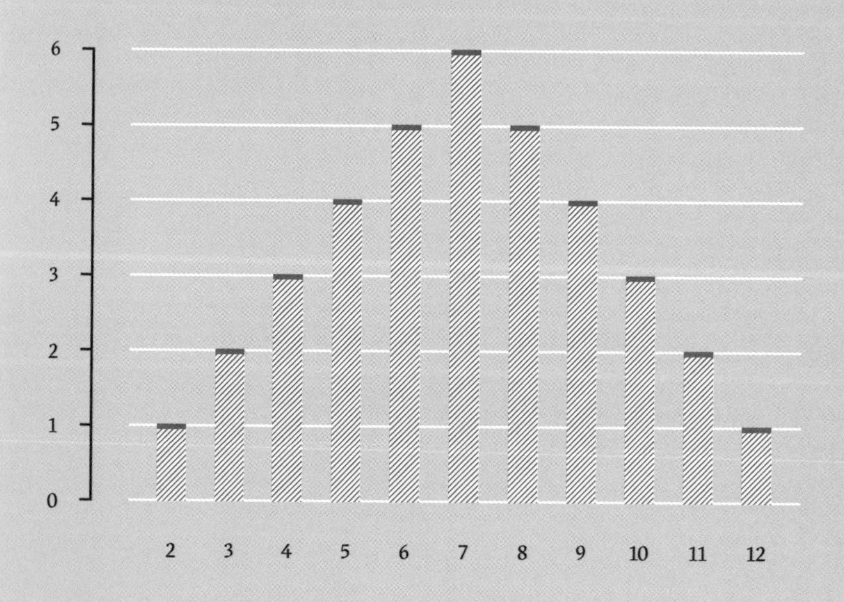

as "standard deviation," which describes the shape of the distribution, enabling us to predict what percentage of the individuals in our sample fall within a particular range.

If we can understand the probabilities and distributions involved it becomes possible to build a mathematical model of the subject being studied. This can't tell you what will happen in the future, any more than knowing that there's a 50 percent chance of tossing a head with a coin will tell you the result of a coin flip. But the model enables us to simulate reality and get a feel for what is likely to happen. Which is excellent if we are a looking at a particularly simple and controlled subset of reality. This kind of prediction works not only for games of chance but for everything, from predicting people's height distribution to the chances of a radioactive particle decaying.

So, it can start to seem that we have a strong handle on the future where randomness is involved—until chaos rears its head.

The turkey's diary

"We are all convinced the sun will rise tomorrow. Why? Is this belief a mere blind outcome of past experience, or can it be justified as a reasonable belief?"
Bertrand Russell, 1872–1970

Randomness is predictable
We're starting to see that our understanding of the world around us is based on patterns. It's how we cope with the different circumstances we face every day—and it's the basis for science. Take a trivial example. I buy a bottle of orange juice. It's a new object in my life. I've never had that specific bottle before. But I know how to open it, because it fits the pattern in my mind of "bottles with a screw cap." So, I twist the cap counterclockwise, open the bottle, and get to my juice.

Imagine I had bought a bottle of hipster juice that instead of having a screw cap was tightly sealed with a cork. I could twist the top counterclockwise forever and never get in. Luckily, I have another pattern that deals with bottles with corks (and a corkscrew on my penknife) so I could still open the bottle.

Our scientific viewpoint is primarily based on the method of induction—we observe what happened before and make the often valid assumption that, under the same circumstances, the same things will happen again. This is the case whether we are thinking of something like the Sun rising each morning or the idea that a series of coin tosses will come up with heads half the time and tails half the time. Even randomness has this degree of predictability.

Logic
The systematic study of inference, where a conclusion is derived from the combination of a series of facts or statements.

Note that induction is different from deduction—something that is rarely effective in the real world (or in science). Deduction enables us to use logic to form a conclusion. So, for instance, if I know that "All dogs have four legs" and I have an animal in front of me with three legs, I can deduce that this animal is not a dog. The problem with deduction in the real world is that, while I can make a definitive observation of the number of legs the animal has, the deductive process depends on the ability to make the earlier statement "All dogs have four legs."

To be definitive about this, I need to have examined all the dogs that exist beforehand—which simply won't happen in the real world. So, I have to support my deduction with some induction. Perhaps every dog I've ever seen has four legs, so I assume that they all do. But, of course, there are actually three-legged dogs out there. Deduction is only as good as the underlying assumption, and as long as this depends on induction, we can do no more than say that this is our best current theory. That is how science usually works.

There are exceptions, of course. If I do have the opportunity to verify my initial statement, deduction is reliable. If I have a box of buttons, I can check every one of them and be able to say, for example, "All the buttons in this box are blue." If I am then given a red button I can deduce with certainty that this button did not come out of my box. But real life is rarely like my button box.

Patterns can catch us out

Bertrand Russell, the British philosopher who famously commented on the sun rising each morning, also made an observation of the experiences of poultry on a farm, which is sometimes reshaped in terms of a turkey's diary. Imagine that a turkey kept a diary which showed how good a day it was having. If we plotted these ratings for each day, it might well provide us with a nice distribution and, being mathematically minded, we might try to use that distribution to make predictions about

the future. And then, a few days before Thanksgiving, from the turkey's point of view, its day would go off the scale (in the bad direction) as it was prepared for the table.

Without knowledge of the circumstances, it might seem that the turkey's "bad day" was a totally random event—and certainly without background information it was totally unpredictable. Yet we humans struggle to accept that there can be deviations from our understanding of the world. In the turkey's worldview, based on its previous experience, being trussed and roasted was not a possible outcome. But it happened. Similarly, when the unexpected strikes, we often try to provide an explanation within our current understanding, where it may well be that the cause of a sudden, unpredicted occurrence is outside of our available information and not explainable without changing our view of the world.

←
The turkey's diary
Turkeys might logically infer that each day will be pretty much like the previous one. Until Thanksgiving.

In effect, what happens to the turkey—and happens to us all the time—is that the wrong pattern is being used to understand the world. One way this manifests itself is through superstition. If we see an apparent pattern linking one occurrence and another, we assume there to be a causal link. The superstition we associate with mirrors, ladders, or black cats is a near-inevitable outcome of making use of induction, because it can be difficult to separate real causes and events that just happen to coincide in space or time.

Often, like the turkey, we have a limited understanding of what is going on. It is then easy to apply patterns that have been successful in the past for other requirements to situations where they just don't fit. It could be that something is genuinely random, but we expect a different kind of pattern to apply. If I toss a coin and get nine heads in a row, it's hard not to expect it to be more likely that the next toss will be a tail because in my mind there's a pattern that tells me half the tosses should be heads and half tails. But in reality the coin has no memory, it doesn't know what has happened before, so there is still a 50:50 chance of getting a head or a tail.

The same thing is seen with sports records. When a team has a run of luck or a player is described as having a "hot hand," we are applying a pattern that implies some underlying causation to what can be a totally random set of circumstances that certainly will have no effect on what will happen in the future.

Predicting what will happen next depends on having a good enough understanding of what is going on. We need to get a feel for what a system is (more on this comes next) and how the nature of that system enables—or prevents—the ability to predict what is likely to happen. And a good starting point for this is to take a look at the humble pendulum.

Playing with pendulums

"If one wishes to make the vibration-time
of one pendulum twice that of another,
he must make its suspension four times as long."
Galileo Galilei, 1564–1642
(trans. Henry Crew and Alfonso de Salvio)

Getting systematic
Whenever we are trying to understand the world around us and the impact of chaos on it, the fundamental unit is the system. In everyday life, a system tends to be a way of referring to a social grouping, often in a negative way ("She spends her time fighting the system"), an approach to doing something ("He has a system for winning"), or a piece of technology ("This sound system is amazing"). But for our purposes, a system has a much wider meaning.

A system simply means a collection of interacting components. It can be as basic as a ball and a slope it can roll down, or as complex as the universe. A pen is a system, as is a smartphone, your body, a country's administration, or the weather. A useful classification is to split systems into two broad kinds. An open system can interact with other elements and systems that are outside the system in question, while a closed system cannot. Most systems in everyday life are open, but to simplify matters, we often treat a system that has limited interaction with its surroundings as closed.

One of the simplest systems is the pendulum, which the great Italian natural philosopher Galileo Galilei spent a considerable amount of time studying in the sixteenth and seventeenth centuries. A basic pendulum consists of an anchor point—say, a hook in the ceiling—a suspension mechanism—it could be a piece of string—and a bob, which is a weight attached to the end of the suspension mechanism.

A pendulum demonstrates the conversion of energy from one form to another. If we start the pendulum swinging from side to side, at the top of the swing, the bob has some potential energy, the energy caused by being lifted up under the force of gravity, but no kinetic energy—the energy of motion. As the bob starts to swing down, some of the potential energy is lost as the bob gets lower and the kinetic energy increases as the bob moves faster. As it moves from side to side, the pendulum is constantly switching energy between potential and kinetic and back again.

Energy
The aspect of nature that makes things happen. As the American physicist Richard Feynman said, "It is important to realize that in physics today, we have no knowledge of what energy *is*."

This is not a closed system. A small amount of the energy initially given to the system by lifting the bob will go into distortion of the suspension mechanism, generating heat, while a little more will be lost to air resistance, unless the pendulum is enclosed in a vacuum chamber. Crucially, the system can't be considered closed as without the gravitational pull of Earth, there would be no potential energy to be transferred into kinetic.

Soon after Galileo's work, pendulums started to be used for timekeeping—Christiaan Huygens's first pendulum clock constructed in 1656 made use of Galileo's observation that the time it took for the pendulum to complete its swing depended only on the length of the suspension mechanism. It didn't matter how heavy the weight was, and bigger and small swings were accomplished in the same time.

In reality, this last observation was only true for relatively small movements, but the pendulum was a well-behaved system with an easily predicted motion. It was the absolute opposite of chaos. Yet one small change is all it takes to alter that.

The pendulum goes wild

Think of a pendulum where, rather than a piece of string, the suspension mechanism is a metal rod. We can dispense with the bob: The weight of the rod will do the trick. It makes no difference to the functioning of the pendulum—it still obeys the expected relationship between length of rod and the period of the pendulum's side-to-side motion. Now cut the rod partway along and put in a connector which allows the bottom section to rotate against the top section, then set the pendulum in motion again.

This is only a simple change. We have gone from a single, continuous suspension mechanism to one where it is split into two, freely moving parts. In effect, it has become a double pendulum, with each part of the rod acting as its own weight. It would not be surprising if this change made the pendulum's motion a little less smooth and predictable. But in practice, the thing goes crazy.

Set the double pendulum in motion and soon the bottom section will be spinning wildly, before suddenly jerking and starting to rotate in the opposite direction. The whole system will jump around as if it is being pulled on by a set of unconnected forces. Its behavior is, to all intents and purposes, totally unpredictable. Yet this is no complex system—it is one of the simplest imaginable. That small change of adding a joint (or, if you prefer, attaching a second pendulum to the bottom of the first) has totally destroyed the predictable motion of the system. It has become chaotic.

→

Composite pendulums
Accurate pendulum clocks often feature pendulums with a mix of materials to counteract the tendency of metals to expand and contract with heat. (Engraving by J. Pass, 1809.)

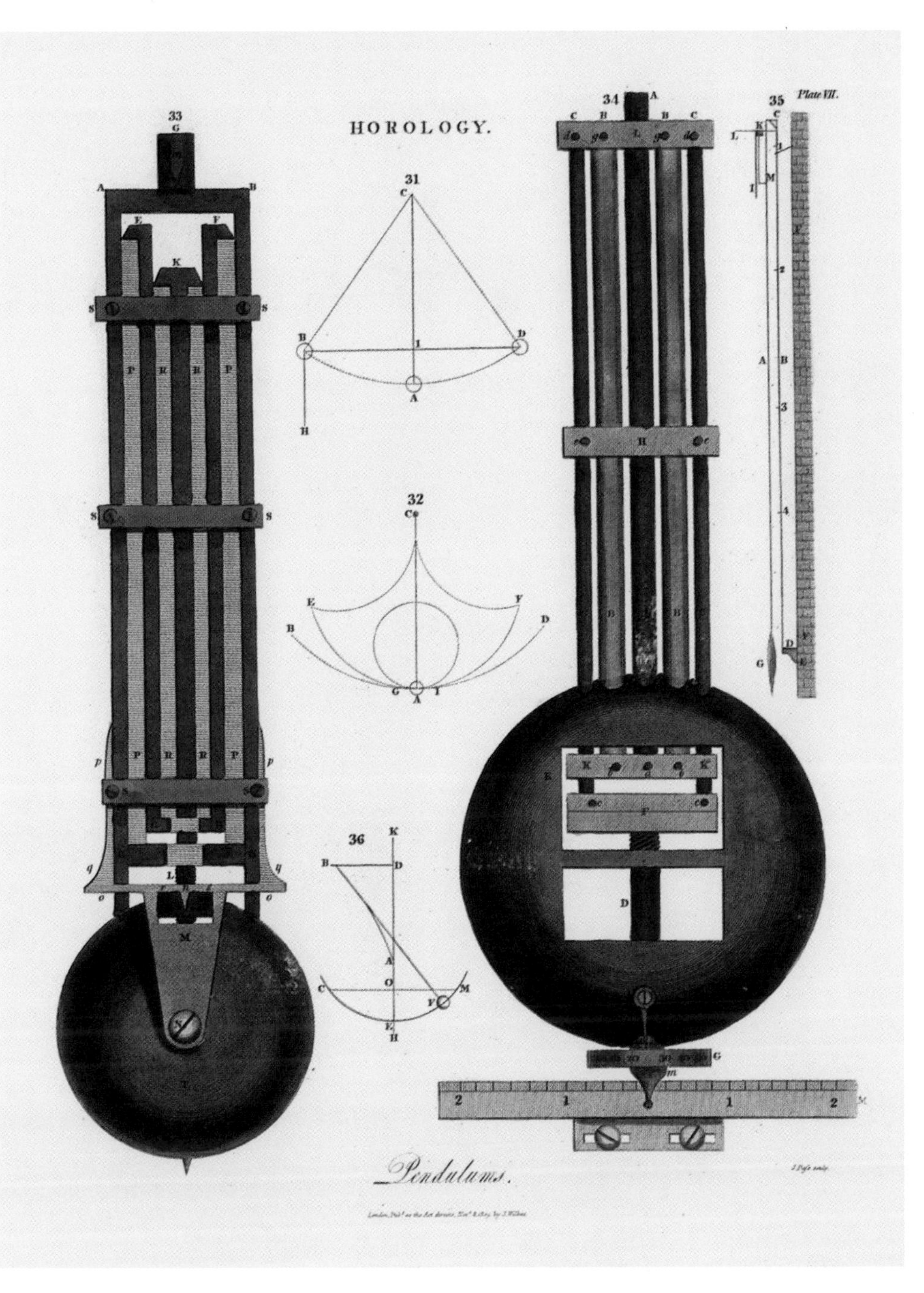
HOROLOGY.
Plate VII.
Pendulums.

Double pendulum path
One example of the motion over time of a pendulum with a single joint. Starting it again would trace a totally different path.

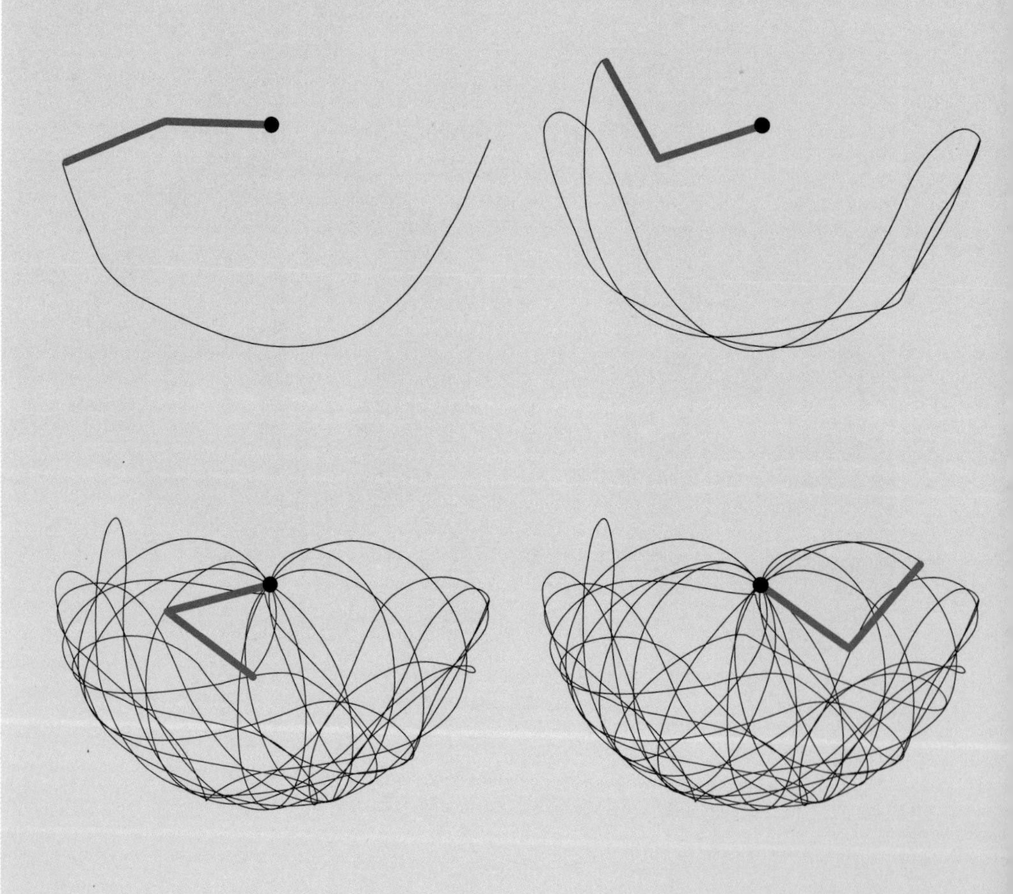

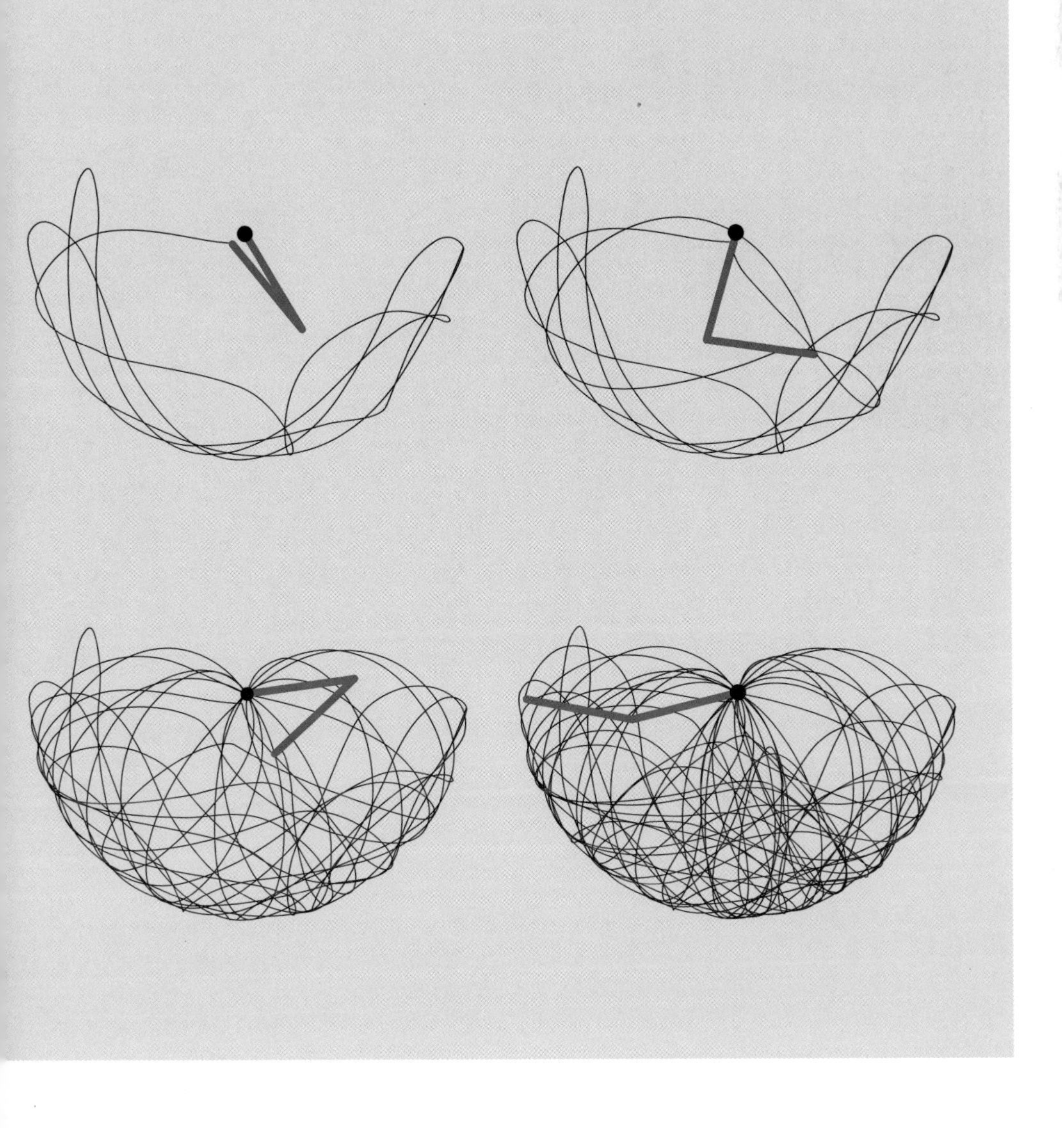

It does not compute

"[All phenomena] are equally susceptible of being calculated, and all that is necessary, to reduce the whole of nature to laws similar to those which Newton discovered with the aid of the calculus, is to have a sufficient number of observations and a mathematics that is complex enough."
Marie Jean Antoine Nicolas Caritat, Marquis de Condorcet (1743–1794)

Hapless computers and deterministic unpredictability
There is something rather touching about the portrayal of evil supercomputers in old movies. These early imaginings of what would now be called artificial intelligence would be set upon an apparently unstoppable path to take over the world or destroy humanity until foiled by a hero who would put the machine into a spin by setting it a challenge that was impossible to achieve. "What is one divided by zero?" the hero might ask. In a more sophisticated movie, he (usually) could come up with a nice bit of twisted circular logic, like "This statement is false. Is that statement true?" The evil computer, unable to help itself, would attempt an impossible mathematical calculation or get into a logic loop. "If the first statement is false, the first statement must be true. But if it's true, then it is false. And if it is false, it must be true." The computer's quandary would go on some time in a higher and higher pitched voice, often

followed by "This does not compute!" before smoke began to emerge from its cabinets, then the whole thing would blow up and the world would be saved.

Despite being great cinema, the computer's worldview, based on rigid logic, was being challenged. The statement it was required to process was not computable. Unlike the Marquis de Condorcet's comfortable position set out in the quote above, all phenomena were not capable of being calculated for our unfortunate computer. And a very similar kind of problem arises from the behavior of the double pendulum. Here we have moving objects obeying Newton's laws of motion—yet the outcome is not subject to calculation.

Specifically, the motion of the double pendulum is deterministic—it should in principle be calculable and it is a very simple system—yet in practice, calculating its motion for very long is beyond us. What we have here is a kind of inverse randomness. A truly random system, say the decay of a radioactive particle, or the toss of our imaginary, entirely fair, coin, has a degree of predictability.

With a chaotic system like the double pendulum, there is no actual randomness. The entire outcome is deterministic. If it were possible to set the pendulum swinging from exactly the same position with exactly the same motion, in exactly the same environment, then exactly the same outcome would ensue. But in practice it is impossible to restart the system with such accuracy, and very small differences in the way the double pendulum starts off make for a totally different sequence of movements. As we will discover, there are many such systems in everyday life, from Wall Street to the weather.

Stormy weather: the definitive chaotic system
The sensitivity of the weather to small differences in conditions inspired the development of chaos theory.

Faking it

Almost all the apparently random numbers we make use of in daily life are, in reality, dependent on a mathematical equivalent of the chaotic system, a pseudo-random number generator. When, for example, we use the random number function of a spreadsheet such as Microsoft's Excel, it does not come up with a truly random number, rather it makes use of a deterministic sequence that jumps around in a near-random fashion.

Although it is not always obvious when used, a pseudo-random number generator starts from a "seed" value—a starting point provided by the computer, which may well be the number of seconds or milliseconds since some fixed date in the past. This starting point is a value that is constantly changing and is then processed with a simple mathematical formula to produce the next item in a sequence of pseudo-random numbers.

A common mechanism called a Lehmer random number generator, which is often used in spreadsheets, generates a value by multiplying the previous value by a specially selected

←

Lottery machine
A familiar pseudo-random number generator, the lottery machine is a complex chaotic system.

large number, then taking the modulus (the remainder) of the result when divided by a large prime number. The seed is fed in first, the outcome of the first calculation is used as the starting point for the same formula in the next iteration, and so on.

There are true random value techniques that make use of quantum processes, either indirectly (for example in the thermal noise in electronic devices) or directly by using a quantum device such as a beam splitter which randomly sends a photon in one direction or another, to generate true random numbers. This technique of using light is how random numbers are generated by the affectionately named ERNIE (Electronic Random Number Indicator Equipment), the device used by the U.K. government's saving scheme called premium bonds. This system is effectively a lottery where once a premium bond is purchased it remains in play until it is sold back at the original price, with the lure of potential winnings from a monthly draw acting as interest on the original investment. The winning numbers are generated by ERNIE using a true random process.

By contrast, most conventional lotteries use a chaotic, pseudo-random system, for example the interaction of a number of balls in a rotating chamber. The balls bounce off each other and the rotating paddles many times before some of the balls are allowed to escape. Although this is a much more complex system than a double pendulum or a pseudo-random number generator it is similar in being deterministic but unpredictable.

The double pendulum is one of the simplest systems to exhibit chaotic nature, but it was not the first to be discovered.
For that, we have to go back to Newton and his law of gravity.

2
Newton's Intractable Motions and Runaway Feedback

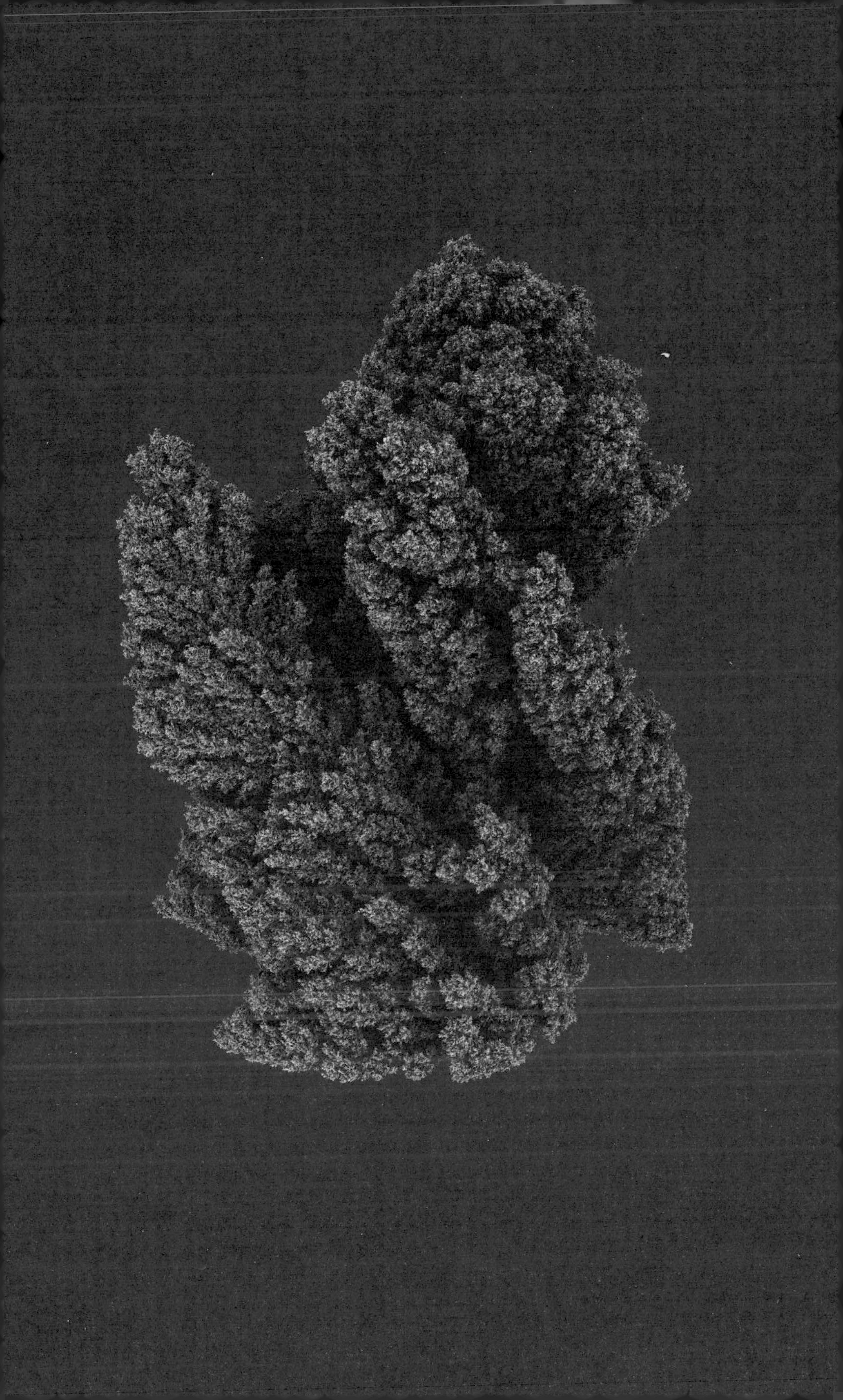

The gravity of the situation

"And thus Nature will be very conformable to her self and very simple, performing all the great Motions of the heavenly Bodies by the Attraction of Gravity, which intercedes those Bodies. ..."
Isaac Newton, 1643–1727

A simple relationship
It might seem that Isaac Newton's work on gravity and planetary motion is the last place we would expect to find unpredictable chaotic behavior. We know that Laplace used Newton's work as the basis for his view of a universe where everything that happens can be predicted from the state of the universe the moment before. And Newton's *Principia* provides a mathematical relationship that describes the force of gravity. Yet Newton was obliged to recognize an oddity when gravitational forces are calculated.

The *Principia* is not an easy book to read. It was written in Latin at a time when many scholars were moving away from science's former universal language to write science books in their native tongues. Newton, however, deliberately made much of *Principia*'s content difficult to understand to ensure its audience was restricted to the cognoscenti. Oddly, he also made things seem more complex in an attempt to make the text more approachable to mathematicians of the day. They were steeped in geometry, so Newton translated much of the work he had undertaken using the newly invented calculus (his "method of fluxions") into far more convoluted geometric forms.

However, some aspects of the book shine through. One is his introduction of the three laws of motion, often now known simply as Newton's laws. But the gravitational relationship between two bodies—whether planets or grains of sand—is less obviously stated. A modern textbook would give the relationship discovered by Newton as something like this:

$$F = \frac{Gm_1m_2}{r^2}$$

Here F is the force of attraction between two bodies, m_1 the mass of the first body, m_2 the mass of the second body, and r the distance between them. G is the gravitational constant, that is a constant value assumed to hold throughout the universe, whatever objects and distances are involved. By implication, this equation lay behind what Newton describes, but in practice he typically talked only of, say, the gravitational force being proportional to the masses involved.

One thing worthy of note here is that, crucial to understanding the force of gravity, Newton invented the concept of mass, as opposed to the more familiar idea of weight. Mass is effectively the amount of "stuff" in an object, which provides its resistance to being accelerated, whereas weight is actually a force—the force felt by an object of a particular mass under the attraction of gravity. Mass is an absolute property of an object—it is the same wherever the object is placed, whereas weight depends on the strength of local gravity. For example, your weight on the Moon, where gravity is far weaker, would be around one-sixth of what it is on Earth.

Mass is measured in kilograms (the U.S. unit of the pound is derived from the kilogram and so is also technically a measure of mass). Strictly, as weight is the force due to gravity, on Earth the weight of an object with a 1kg mass is 9.81 newtons.

In practice, though, we cheat and refer to the mass as if it were a weight. This only gets confusing away from Earth—so on the Moon, that object still has 1kg mass, but only the equivalent of 0.17kg weight.

By comparison with the complexity of the mathematics of much modern physics, the relationship that gravitational force has with mass and distance is beautifully simple. So, in some ways, it is quite surprising that the equation, or something like it, never appears in the *Principia*. One reason was that the gravitational constant had yet to be defined. Its value was first implied when English scientist Henry Cavendish made a successful attempt to calculate the mass of Earth in 1798 and the equation as set out above was, remarkably, only written down in the 1890s, when another English physicist, Charles Boys, specifically gave a value for the constant.

All that Newton acknowledged was that the attractive force of gravity was proportional to each of the masses and inversely proportional to the square of the distance between them (or, to be precise, the distance between their centers of gravity, as Newton also showed that two bodies could be treated as points, located where the gravitational pull of the matter around the point balanced out).

Earth and the Moon

In the *Principia*, Newton looks at a range of astronomical orbits, but perhaps the simplest and most elegant to be covered is his analysis of the relationship between Earth and the Moon. The Moon orbits Earth (or again, to be accurate, the Moon *and* Earth both orbit a point between them, called the barycenter, but Earth is so much more massive than the Moon that this point lies around 1,000 miles (1,700 kilometers) beneath the surface of Earth.) When one object orbits another it is effectively doing two things: falling toward that body and missing it.

An orbiting satellite (here, the Moon) is constantly falling toward Earth, accelerated by the force of gravity. But to stay in orbit, it is also moving sideways at 90 degrees to that fall, with sufficient speed to stay the same distance away from Earth. This, incidentally, is why astronauts on the International Space Station (ISS) float around in pretty much zero gravity. They are really not far from the surface of Earth—just 220 miles (350 kilometers) up. The force of gravity at that distance is around 90 percent as strong as on the surface. But the station and the

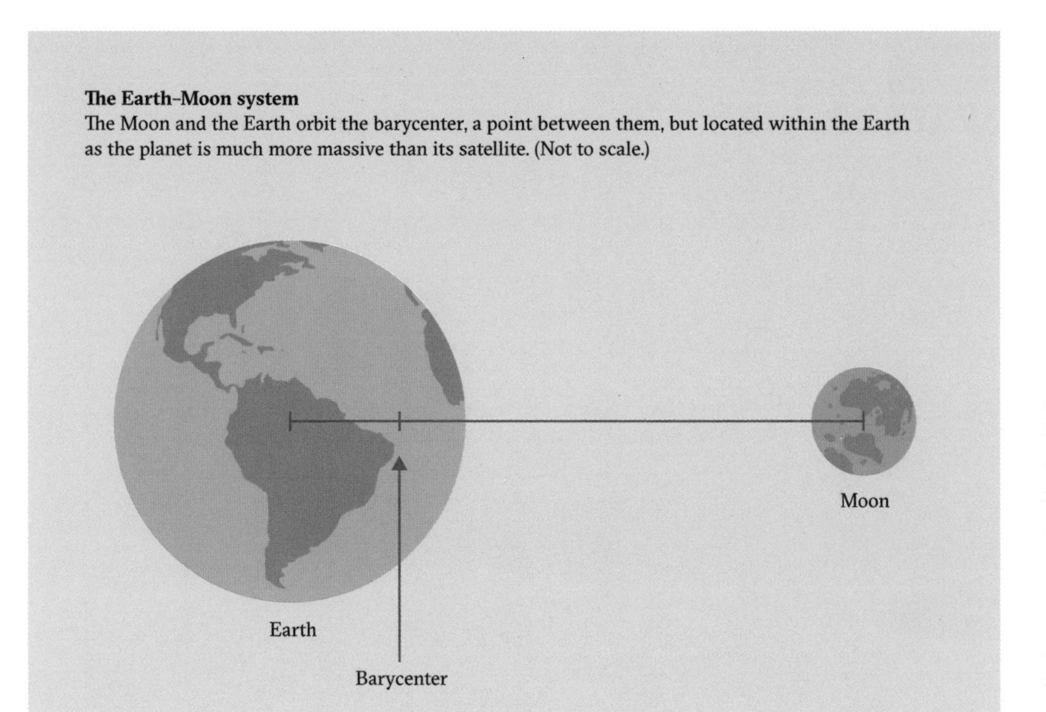

The Earth-Moon system
The Moon and the Earth orbit the barycenter, a point between them, but located within the Earth as the planet is much more massive than its satellite. (Not to scale.)

astronauts are in free fall toward Earth so they feel no force; they float around as they would in a plummeting airplane or elevator. Luckily, though, the ISS is also traveling sideways at just the right speed to maintain that height.

Given the best astronomical measurements of his day, Newton was able to calculate the rate at which the Moon was accelerating toward Earth. We know he didn't have the above formula for the force of gravity, but we can much more easily make the calculation by combining it with Newton's second law of motion, that the force applied to a body is its mass times the acceleration it produces. So we can say that:

$$m_{moon}a = \frac{Gm_{\text{moon}}m_{\text{earth}}}{r^2}$$

This means that the acceleration is just

$$a = \frac{Gm_{\text{earth}}}{r^2}$$

International Space Station
The space station is in free fall toward Earth—but its sideways motion keeps it at the same altitude.

→

Newton's cannon
The diagram shows the theoretical paths of projectiles from an imaginary 5-million foot-high (1,524-kilometer-high) mountain on a simplified Earth.

Note that the acceleration is not affected by the mass of the object being accelerated—the mass of the Moon is canceled out. In a simplified English version of the third book of the *Principia*, named *Of the System of the World*, Newton considered what would happen if a cannon on a mountaintop fired a projectile sideways at increasing speeds until it continued right around Earth, effectively orbiting at Earth's surface. This feels a little bizarre, because we're used to orbits being up in space. In principle it's possible, but in practice it's messy, partly because Earth doesn't have a smooth surface, so the projectile would have to avoid the bumpy bits (hence the mountain firing site), and partly because the speed required to keep missing is very high—near the surface it would be around 17,600 miles per hour (25,000 kilometers per hour).

Newton calculated the acceleration that a body (such as the Moon) would encounter if it were in fact orbiting at practically no height above Earth's surface. This proved to be pretty much identical to the acceleration due to gravity that we feel on Earth's surface—32 feet per second per second (9.81 meters per second per second).

Newton had shown with an impressive feat of mathematics that the same force of gravity was responsible for the familiar fall of things we drop on the surface of Earth and the motion of the Moon in its orbit. With extra calculations on, for example, the moons of Jupiter, he demonstrated that his gravitational theory appeared to be literally universal—it applied everywhere in the known universe. Two apparently very different things were in fact the result of that same simple relationship.

In reality, the precise details of the gravitational relationship of Earth and the Moon are not simple, because neither is a perfect sphere, and both are pulled around by gravity, resulting in not only the familiar tides of the oceans but also a small tidal movement in the land itself. However, what Newton had achieved in his mathematical account was a mechanism that, given the correct values, could be totally precise in its description of the way that the force of gravity allowed two bodies—such as Earth and the Moon—to interact with each other.

However, things are more complicated than simply having a system consisting of Earth and the Moon. We live in a far messier universe. For as long as people have gazed into the night sky they have been aware of greater complexity.

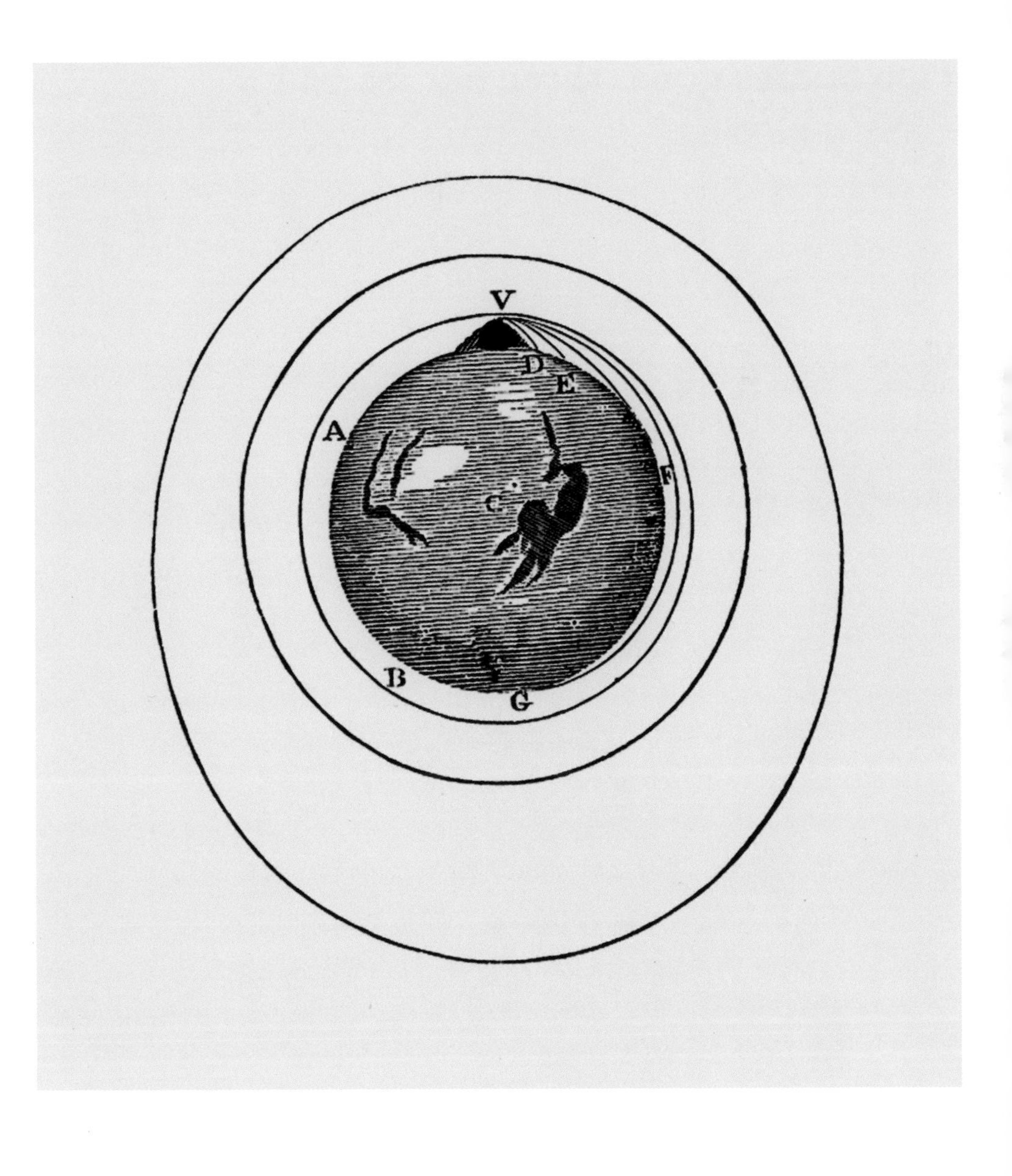
V
D
E
A
F
C
B
G

Two bodies good, three bodies bad

"There is no one center of all the celestial circles or spheres. The center of the Earth is not the center of the universe, but only of gravity and of the lunar sphere."
Nicolaus Copernicus, 1473–1543

The power of attraction
Newton had moved the idea of gravity on from the simplistic picture that Copernicus's words here suggest, where gravity is only related to how things fall on Earth. It took Newton's insight to show that gravity was the linking force between heavenly bodies, and that dispensed with the need for celestial spheres. In the process, though, he opened the door to chaos.

Gravity has no limits. This applies in two different ways. First, there is no material capable of stopping gravity. The attractive force passes through solid objects as if they were not there. There can never be a material like the cavorite that features in H. G. Wells's novel *The First Men in the Moon* (1901): an imaginary substance that blocks the force of gravity. If there were such a thing, we could get energy from nowhere. Think of a waterwheel with a series of metal paddles. If we were able to paint one side of each blade with our gravitational blocker, one side of each paddle would be attracted toward Earth, but the other side would not. The wheel would turn of its own accord. The side of the wheel where the untreated parts of the paddles faced downward would be pulled toward Earth's surface, with no counterforce on the other side. While the idea is an attractive one, all the evidence is that such a shattering

Laws of Thermodynamics Four laws of nature that determine how heat and other forms of energy move within systems.

of the first law of thermodynamics—the conservation of energy —is totally impossible.

However, there is another way in which gravity is unbounded. No *distance* stops the attractive force of gravity either. Admittedly, the force of gravity rapidly gets weaker with the square of the distance. As the r on the bottom of the equation on page 51 gets smaller, the force drops away. Double the distance and the force reduces to one quarter of its value. But, however far you separate two bodies, the attraction between them never reaches zero. Every single object in the universe has a gravitational influence on every other. In this respect (and only this respect) astrology makes a kind of sense—because the position of the planets does have a gravitational effect on you as an individual. Unfortunately, astrology is still a fiction, in part because there is no reason why the gravitational influence of the planets at your birth should change anything, and also because the planets are so far away that the other people in the room when you are born will have a bigger gravitational influence than these distant objects do.

Nonetheless, everything in the solar system affects everything else gravitationally—and nothing is more impossible to ignore than the Sun. It's easy to underestimate the Sun, because, viewed from Earth, it looks about the same size as the Moon. We know it is bigger, but it's difficult to assimilate the idea that it's 400 times the Moon's diameter. The Sun contains about 99.9 percent of the entire mass of the solar system. As a result, despite its vast distance—averaging around 93 million miles (147 million kilometers) from Earth—it has a significant effect on the system of Earth and Moon. It's not realistic to consider these two bodies in isolation—there's a third party in this relationship. Newton was aware of this—but he also knew that his precise calculations went out of the window as soon as the influence of the Sun was included.

The problem is that once a third body is involved, it's not just a matter of Earth's influence on the Moon and the Moon's influence on Earth. The Sun will impact each of these—and for that matter, each of these will influence the Sun's effect on the other. Just like adding the joint to the pendulum, adding a third body to the two-body gravitational problem injects chaos into previously sedate proceedings. Although in principle it should be possible to calculate exactly what will happen, in practice it is infeasible. Newton couldn't do it. And this wasn't down to a lack of serious computing power in the seventeenth century. We can't do it either.

Although the resultant motion was neither named nor understood in the same way as it is now, Newton had established that the universe was not going to give up all its secrets easily, not even to the power of mathematics. With more than two bodies, chaos reigns.

Perturbation to the rescue

Newton would not allow this inability to undertake a precise calculation to get in the way of demonstrating his genius in the *Principia*. To cope with the effects of the Sun (in fact he generalized his approach to any such combination of bodies), he realized that he could make use of a principle known as perturbation. The idea is that, rather than calculate the exact relationship of the three interacting bodies, you first calculate the outcome for the two bodies that interest you, then consider those as one unit to see how a change caused by the third body is likely to affect them.

We're used to physics being about exact values, which is usually achieved through extreme simplification of the situation. Here, it is accepted that the exact values cannot be achieved, and perturbation provides us with an approximate result. Although it can never be exact, the better the calculation, the closer it will be to the correct value.

So, for example, Newton's gravitational equation can be used to calculate the force of attraction between Earth and the Moon, predicting the way that the Moon and Earth orbit each other in isolation. Newton then calculated the forces that the Sun would have on the Moon and Earth in their solo orbit, and what effect the gravitational force of the Sun would have in changing that orbit. The outcome was not perfect—it assumed Earth and the Moon had no impact on the Sun—but

→
Newton's perturbation diagram
In the *Principia*, Newton shows how the influence of the Sun on the Moon's motion can be approximated.

given how much bigger the Sun is than either of them, this wasn't significant.

Not surprisingly, adding in even more bodies makes the problem trickier still. For much of the solar system, for example, the planet Jupiter also has a significant gravitational impact. Again, it is possible to break down the problem by looking at pairs of bodies and adding in perturbation, but the calculations required become increasingly complex and the approximations need more careful handling. Nowhere is this clearer than when dealing with a whole galaxy at a time.

Prop. XXV. Prob. V.

Invenire vires Solis ad perturbandos motus Lunæ.

Deſignet *Q* Solem, *S* Terram, *P* Lunam, *PADB* orbem Lunæ. In *QP* capiatur *QK* æqualis *QS*; ſitque *QL* ad *QK* in duplicata ratione *QK* ad *QP*, & ipſi *PS* agatur parallela *LM*; & ſi gravitas acceleratrix Terræ in Solem exponatur per diſtantiam *QS* vel *QK*, erit *QL* gravitas acceleratrix Lunæ in Solem. Ea componitur ex partibus *QM*, *LM*, quarum *LM* & ipſius *QM* pars *SM* perturbat motum Lunæ, ut in Libri primi Prop. LXVI. & ejus Corollariis expoſitum eſt.

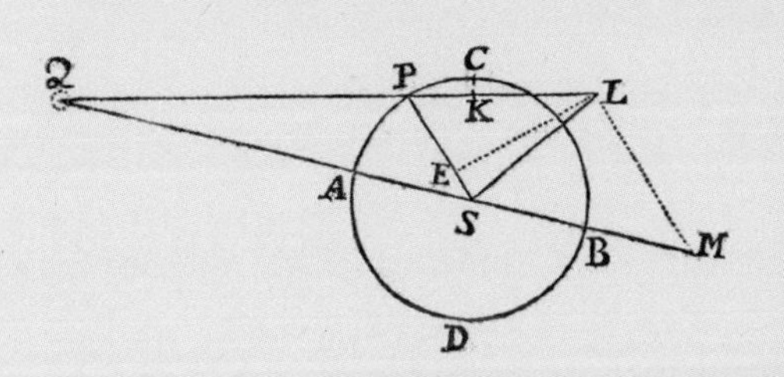

Qua-

Solar influence
Not only does the Sun give us light and heat, its gravitational force keeps Earth in orbit and influences the Earth–Moon system.

Spinning galaxies

"The evidence at present available points strongly to the conclusion that the spirals are individual galaxies, or island universes, comparable with our own galaxy in dimension and in number of component units."
Heber Curtis, 1872–1942

Island universes
Once Newton had broken for good the need for heavenly spheres, our appreciation of the scale of the universe grew beyond what we would now think of as the solar system. It was realized in time that the magnificent swathe of light that forms the Milky Way was a collection of many billions of stars, of which our Sun was just one.

Until the 1920s, however, there was considerable debate over whether there was more to the universe than the Milky Way. There were scientists who thought that fuzzy little patches of light, some of which resolved into elegant spirals when viewed through telescopes, were just local groups of stars or patches of gas. But others, such as American astronomer Heber Curtis, who coined the expression "island universes" while debating with another U.S. astronomer, Harlow Shapley, believed—as we now know—that each of these tiny patches of light were huge collections of stars, what we now call galaxies, like the Milky Way.

Galaxies tend to spin. In fact, pretty much everything in the universe spins. This motion reflects the way that the structures formed from clouds of gas or dust: Unless the cloud was exactly symmetrical (which it was never going to be) there would be more gravitational pull from one side than another, and as the material condensed it would start to spin. But by the 1970s, something strange was observed. Building on the work of Swiss astrophysicist Fritz Zwicky, American astronomer Vera Rubin noticed something strange. The galaxies were spinning too fast. Why was this?

The potter's wheel effect

Spin a lump of clay on a potter's wheel too fast, and pieces of the clay start to fly off. The force of the adhesion between bits of clay is eventually less than the force that sends the clay flying off. The same thing should happen with galaxies: If they rotate fast enough, the gravitational attraction should be too small to hold the galaxy together. And when Rubin measured the speed of rotation of a number of galaxies, they were spinning sufficiently quickly that stars should have been shooting out of them like sparks from a pinwheel.

There seemed only one possibility, something that Zwicky had named in the 1930s but that wouldn't be taken seriously until Rubin's discovery, forty years on. There was extra matter in the galaxies that was invisible. Zwicky had called this *dunkle Materie* in German—in English this became "dark matter." Remarkably, calculations suggested that there was five times as much of this invisible matter, interacting only gravitationally, than ordinary matter.

Dark matter remains a mystery. Despite many experiments dedicated to finding it, there has never been a single particle of dark matter detected directly. It is only indirectly apparent from its gravitational effect. Some physicists have proposed that dark matter does not exist as a substance—that the effect is caused by a subtle variation in the effects of gravity on the scale of a galaxy. But an American mathematician, Donald Saari, has suggested that there is nothing to explain, that the cause of the apparent effect is a massive expansion of the problem Newton faced in dealing with the gravitational combination of Earth, the Moon, and the Sun.

The Milky Way
Shown over the Pedra Azul peak in Domingos Martins, Brazil, our galaxy, the Milky Way, is visible in dark skies as a ribbon of light.

An uncomfortable darkness

Saari points out the difficulty of making a gravitational calculation based on not a handful of bodies in the solar system but the billions of massive stars in a galaxy. According to Saari, the calculation to determine how fast a galaxy should be able to rotate without flying apart is fatally flawed.

Single body
Newton showed that a collection of parts such as Earth acts as if it were a single object at the center of gravity. This assumes, however, that the parts aren't moving with respect to each other.

Clearly, it's not possible to calculate the precise interaction of each of the stars (bearing in mind the maximum number of bodies for total accuracy is two). So, astronomers made some significant approximations. Take, for example, the beautiful Andromeda galaxy, our nearest large neighbor, which contains around a trillion stars, and measures some 220,000 light years across, where a light year is around 5,880,000,000,000 miles (9,460,000,000,000 kilometers). It's a mind-boggling task to work out what's going on.

The way astronomers approached their task was to look at a specific star and consider the entirety of all other stars closer to the galaxy's center as a single body, so there was just a two-body relationship to consider. (The stars farther out can be ignored as they roughly cancel each other out.) But the reality of the interaction is not that simple. The attraction between a star and its nearest neighbors introduces a chaotic element that overwhelms the impact of the rest of that imagined continuum—typically, for example, a faster neighboring star will tend to pull our target star along with it. It seems possible that the problem that fazed Newton has, on a large scale, deceived modern astronomers too.

In their spare time, though, the astronomers may play a game where chaotic effects can once more be seen disrupting Newtonian neatness.

←
The Andromeda galaxy
The nearest large neighbor to the Milky Way, the Andromeda galaxy is around 2.5 million light years distant.

Billiards versus pinball

"Or if you try to 'see' an electron with light (or X-rays), then the light photons strike the electron like billiard balls. ..."
Roger Jones, b. 1953

The physicist's favorite game

Before we start to get a feel for how this unpredictable behavior arises, one other example from Newton's world is worth exploring as a hint of the nature of chaos to come: games where a ball bounces off other objects. One of the earliest such games, and one that would become a favorite example to illustrate physics problems based on Newton's laws, was billiards. Similar to pool or snooker but played with only three balls (two white and one red), billiards originated as a lawn game in fifteenth-century Europe and was already becoming popular in Newton's day. For comparison, pool and snooker were only dreamed up in the nineteenth century.

Generally a game of billiards only involves a single ball striking another ball at any one time. Players hit their own white cue ball with the end of the cue and that ball rolls across the table and (should) hit the red ball. On its course, it might bounce off the sides of the table, known as cushions, adding to the fun for those who set physics problems. This makes billiards ideal for applying Newton's laws of motion. Knowing the position and the momentum of the balls, it should be possible to predict pretty well exactly what the outcome will be when one ball hits another. Strictly speaking, this is only true for the physicist's idealized table where there is no friction as the ball moves over the green baize, and no loss of energy to noise and heat when balls hit each other or a cushion. But in practice, the real-world billiards table, with a reasonably good surface, is a simple enough environment to make the predictions of physics accurate. A good billiards player is reliably capable of pulling off a desired shot. The billiards table has become a showcase of mathematical determinism in the physics world.

Quantum billiards

To be precise, the deterministic nature of billiards was true when dealing with actual billiard balls and tables. However, scientists work with models—mathematical analogies used to describe the behavior of something that appears to resemble a known system. With the development of quantum physics during the 1920s, it became clear that modeling the behavior of quantum particles, such as electrons, photons of light, and atoms, as if they were billiard balls had serious flaws.

Take a simple billiards move—bouncing a ball off the cushion. The ball behaves in a classic, predictably Newtonian fashion. It approaches the cushion in a straight line at a particular angle, hits the cushion and bounces off to travel away at exactly the same angle, but in the opposite direction. It seems reasonable at first sight that, say, photons of light do exactly the same. The photons bounce off a mirror (actually each photon is absorbed by an electron in one of the mirror's atoms, then re-emitted, but let's not worry about that), emerging at the same angle as they arrived, but in the opposite direction. As we were taught at school, "The angle of incidence is equal to the angle of reflection." But, according to quantum physics, this doesn't happen. In fact, the photon takes *every* possible path on its way to the mirror and reflects off at every possible path as it flies out—each path having a different probability. In practice, a lot of these probabilistic paths cancel out, resulting in what appears to be a conventional billiard-ball bounce. However, if we remove some strips of reflecting material from the mirror,

Conventional reflection
Classical physics assumes that light reflects off at the same angle as it arrives, like a billiard ball bouncing off the cushion.

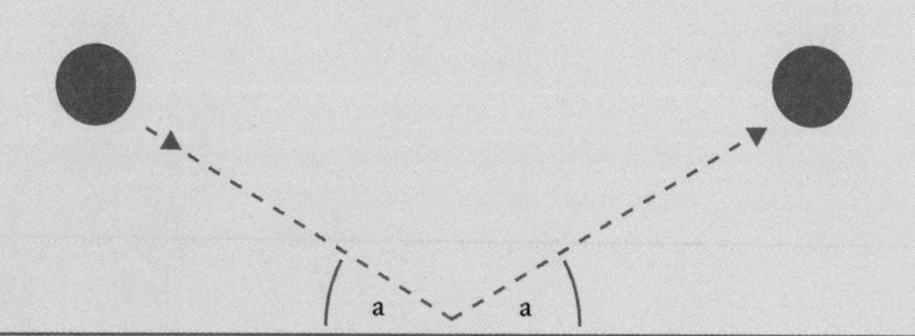

Quantum reflection
Quantum physics shows that photons reflect at many angles—most cancel out, but by removing strips of a mirror, the observed reflection will be at an unexpected angle.

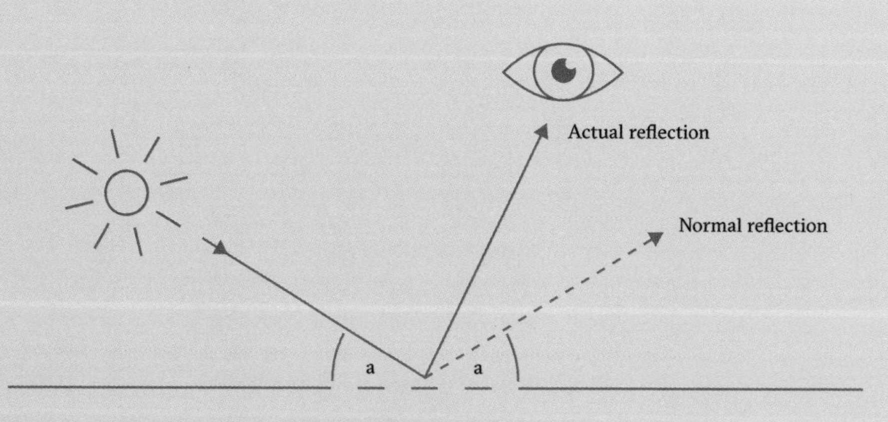

→
CD rainbow
The rainbow patterns are generated by different colors reflecting at unexpected angles.

meaning some of the paths that had canceled out are missing, the light will bounce off at a totally unexpected angle.

The resultant angle depends on the distribution of the missing strips and the energy of the photons involved (the color of the light). It's possible to see this effect in action with a humble CD, which over its playing surface has a large number of pits. These tiny indentations act as if they were missing parts of the overall reflective surface. Tilt a CD at an angle to the light and you will see little rainbow patterns forming. These color effects are produced by varied energies of photons being sent off in directions entirely different from the angle of incidence.

Pinball wizard

The behavior of quantum particles is truly random, but is governed by probability. However, make a few modifications to a billiards-style interaction and another example of the difficulties of making calculations in the real world comes into view. Once again, what is happening is that a very simple interaction becomes impossible to predict because the interplay of factors that make very small differences in the starting point result in very large differences in the outcome—an occurrence that we will come to recognize as an essential outcome of the reign of chaos.

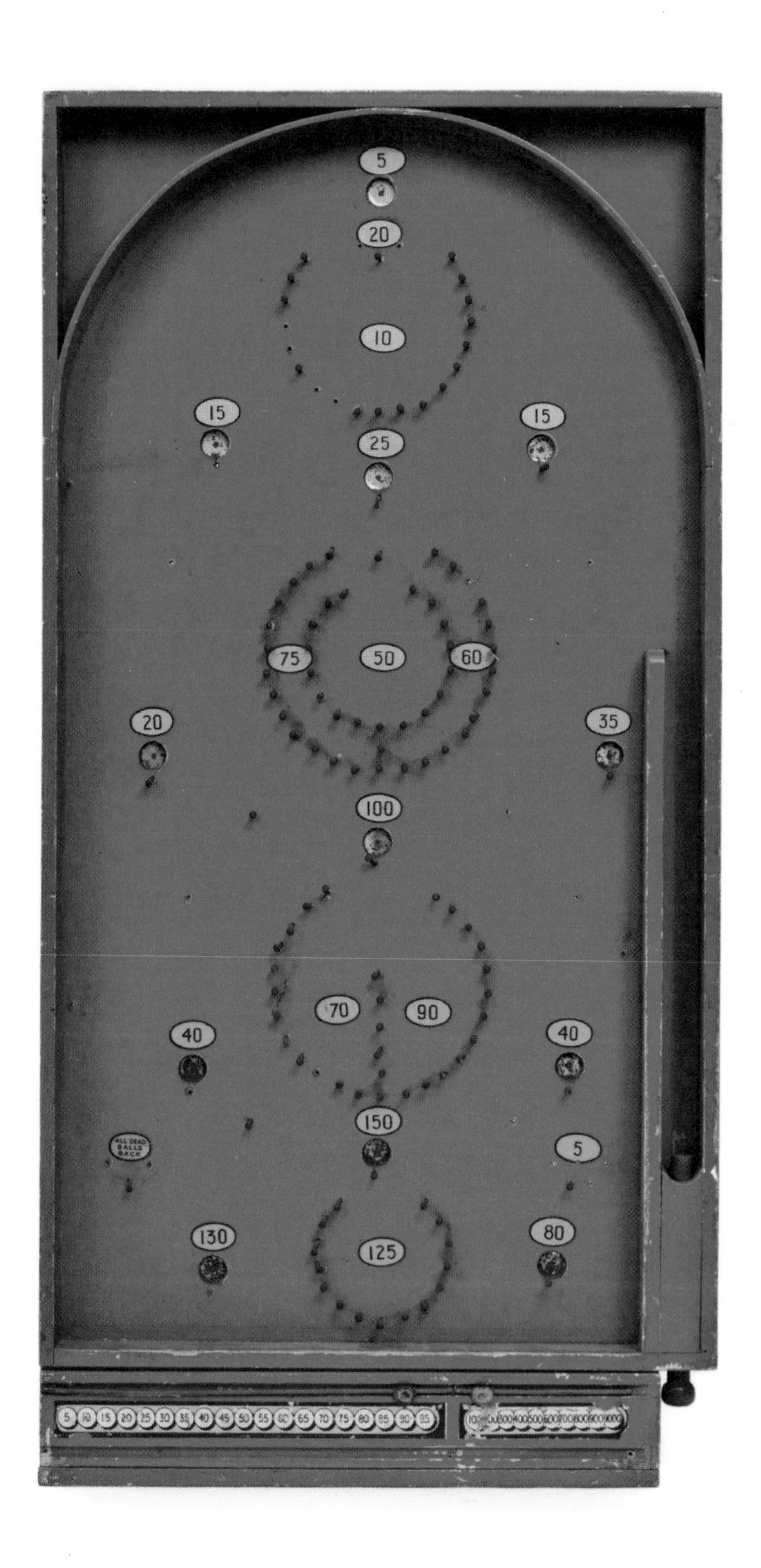
5
20
10
15
15
25
75
50
60
20
35
100
70
90
40
40
150
ALL DEAD BALLS BACK
5
130
125
80
5 10 15 20 25 30 35 40 45 50 55 60 65 70 75 80 85 90 95

This is clear to anyone who has played pinball. While it's true that some are far better than others at playing the game, it is far harder for an expert pinball player to be consistent than is the case with billiards. When a ball is launched into the game, the interaction of the ball with the first obstacle it meets is influenced by several different factors. There is the strength of the push that the ball is given initially—far harder to judge with the spring-loaded device in a pinball machine than with a manually controlled billiard cue. There is any interaction the ball has with the walls of the run up to the top of the game. And should the first obstacle met by the ball be a bumper, the uncertainties in the sensitivity of that component and the thrust the ball will receive are enormous. Similarly, with an old-fashioned pinball table, where the ball collides with a simple pin sticking out of the table surface, the angle of deflection the ball is given is so dependent on very tiny differences in the striking direction that the outcome is unpredictable.

By the time the ball has progressed down the table, its final resting place is beyond realistic control. While a good billiards player can strike the red ball with a cue ball and get it reliably into a particular pocket, no pinball player is able to consistently launch a ball in such a way that it ends up at the same place on the table each time without further interactions from flippers. No matter how carefully the ball is released, the outcome is likely to be different. The billiard table is Newtonian in its determinism. Although the pinball table (unlike the quantum collision) is also deterministic, the outcome is unpredictably chaotic.

Here the chaos is derived from the complex interplay of factors that impact the route of the ball. But in other circumstances, it can be the way that a particular factor feeds back on itself that introduces the interplay necessary to induce an unexpected effect—an effect that it is hoped can be avoided when using a governor.

←
Bagatelle game
The predecessor of pinball and pachinko machines, where striking a pin causes an unpredictable change of direction.

Steam safety

"A governor is a part of a machine by means of which the velocity of the machine is kept nearly uniform, notwithstanding variations in the driving power or the resistance."
James Clerk Maxwell, 1831–1879

The governor's job
Early steam engines were dangerous things. They were developed significantly earlier than the science required to understand fully what was happening to make them work. Explosions and out-of-control runaways were not unusual. This problem, though, had occurred to Scottish steam pioneer James Watt in the eighteenth century. He would solve it with what became known as a governor: not the political position, but a mechanical device.

Watt's governor consisted of a pair of weights attached to hinged bars on either side of a rotating shaft powered by the engine. As the shaft turned faster, the weights flew outward. This mechanism was then linked to a valve controlling the available steam pressure, decreasing the power given out if the weights flew out too far. The result was that the rotation speed of the shaft was controlled or "governed."

Other governors, we learn reading from Scottish physicist James Clerk Maxwell's survey of their mechanism produced in 1868, allow the weights to press against a surface, causing friction. Maxwell's name for these devices, and Watt's version, was "moderator" rather than governor, because it was still

Simple governor
As the shaft rotates faster, the balls fly outward due to centrifugal force, closing the valve.

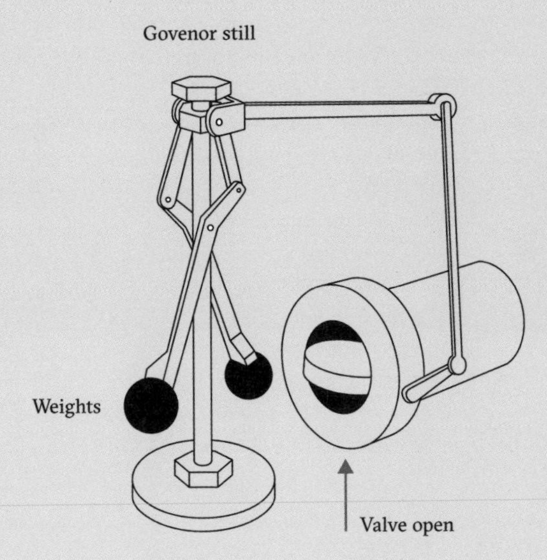

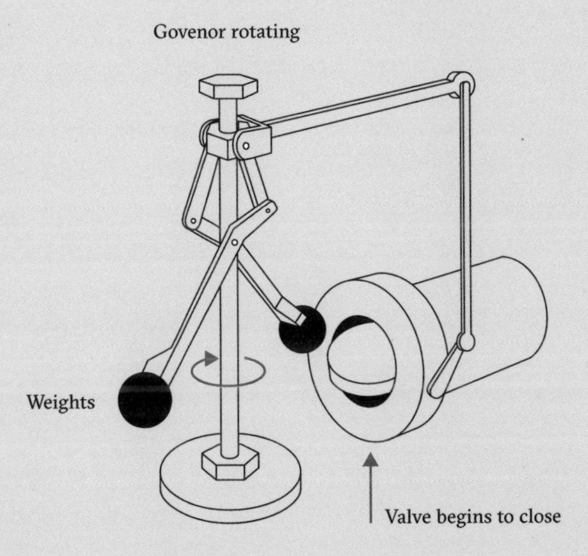

possible to increase the speed by piling on the power. The degree of increase was smaller than it would otherwise be, but it wasn't reversed. The moderator acts as a brake that increases its resistance as the speed increases.

However, Maxwell also came up with an alternative concept, a true governor to his mind, which did not just slow things down as the speed got too high, but actively increased the amount of resistance when above the desired speed and similarly increased power when below the desired speed. It allowed a more sophisticated control mechanism that homed in on the intended outcome.

A familiar and simple example of such a governor is the thermostat, which turns on our heating if the temperature falls too low and turns it off if conditions get too hot. More subtly, the cruise control fitted to some cars, which has more flexibility than simply "off" or "on" and actively returns the car to the desired speed as conditions change, is a modern example of a governor.

In his paper, Maxwell goes into detail of the mathematics of how a governor operates, but absent from it is one essential word, which provides us a link with chaos. This word is "feedback."

Feedback
The return of part of an output signal or flow to the input to increase or decrease the output.

The power of feedback

Simple feedback mechanisms actually date back more than 2,000 years in the form of the float valve, a variant of which is still in use in many modern toilets to shut off the water to the cistern. The idea is to keep the level of a liquid in a tank at a constant by linking a float that moves up and down with the liquid to a valve which allows more liquid into the tank. This means that the result of the level dropping is that more liquid flows in until the float rises high enough to shut off the flow. Usually the "overshoot" mechanism here is an overflow—if the water rises too high it flows back out of the tank through an appropriately sited pipe.

Although crude centrifugal governors (to be less fussy about the term than Maxwell) were used in windmills from the seventeenth century, Watt seems to have been the first to have adopted the word "governor" in the 1780s. But feedback is a far more modern term, first appearing (as feed-back) in the first decade of the twentieth century. Feedback describes what the governor does, producing an effect on the machine that is dependent on the size of the output (speed, for instance), and which changes the value of that output.

Feedback comes in two flavors. The governor is indulging in negative feedback, which confusingly tends to have the more positive action. In a steam governor, the greater the speed, the more the governor shuts it down. The governor's action operates in the opposite direction to the input that triggers it, here the speed of rotation of the engine. Negative feedback calms things down—which is clearly the point of having a governor. However, its dark cousin is positive feedback.

Management-speak uses the term "positive feedback" to mean giving someone a verbal pat on the back but the scientific meaning—and the one that applies to systems to this day—is far more risky. A positive feedback makes a change in the *same* direction as the value that triggers it. So, for example, if we had a positive feedback device on our steam engine, if the shaft is rotating too quickly, instead of sending an instruction to slow the engine down, the feedback device will tell it to speed up. Now it will be going faster still, causing the feedback device to push for even more speed. The result is an out-of-control, runaway system.

Screeching speakers

A positive feedback system gets chaotic when, for example, a live microphone is taken near a speaker that is attached to it. We've all been there when someone speaks into a microphone and the speakers screech, but even without a direct user, bringing a microphone near a speaker results in the same ear-piercing sound.

What's happening here is that very small background noises are being picked up by the microphone, amplified, and pushed out through the speaker. The sound from the speaker now adds to the background noise picked up by the microphone, so a louder sound comes out of the speaker. The noise gets louder and higher in pitch as the amplifier struggles to cope. Predicting exactly what noise will come out as a result is impossible—the process is too chaotic—but we can be sure that the result will be ear-piercing in intensity. It tends to be high pitched as a result of a combination of three things: the distance between the microphone and the speakers, directionality, and inconsistent responsiveness of both the microphone and the amplification system to different frequencies.

Responsiveness
Both microphones and amplifiers can be more sensitive to some frequency ranges, resulting in a distortion of the sound produced.

With its connotations of ear-splitting squeals and out-of-control machinery, feedback might seem an entirely negative thing—but bear in mind that, behind the scenes, it is precisely the same process in its negative form that keeps systems under control.

→
Body temperature regulation
Despite being immersed in ice water, this swimmer in Svalbard, Norway, maintains a normal body temperature.

Feedback can also be seen in mathematical processes—the pseudo-random number generator used for lotteries (see page 47) works by feeding back the value generated to seed the next run of the generator—and some of the fractal shapes that come up in chapter 3, such as the Mandelbrot set, rely on feeding values back into the equation. Here, although chaos is generated, the result does not run away uncontrollably but produces an unpredictable series of patterns.

Back to life, back to reality
Feedback for control is common in both mechanical and electronic systems, but evolution also regularly produces it (in fact, evolution by natural selection is itself a form of feedback process). Feedback is essential in all biological systems too. Many living organisms—ourselves included—maintain levels of any number of chemical substances, as well as keeping physical parameters such as temperature within a tight range of possibilities. In charge of all of these are biological feedback mechanisms, often involving homeostasis—literally "keeping still in the same state."

Chemical levels such as blood sugar and calcium are maintained strictly by the body, the most obvious of the external feedback processes to affect us is temperature. Unless a person is ill, body temperature stays consistent to a degree either way, even when ambient temperatures are varying by tens of degrees. Our internal sense monitors, thermoreceptors, keep track of changes in body temperature.

If this temperature falls, blood flow to the skin and limbs is reduced by tightening the blood vessels, a reaction that helps prevent low external temperatures from affecting the body. Internal chemical processes are triggered that generate heat, and, if temperatures still don't stabilize, shivering begins, using muscular effort to generate friction and push up the body temperature. Similarly, if the body temperature rises, a negative feedback effect comes into play by producing sweat, which evaporates from the skin and cools it down.

Such responses are often caused by developments in the weather—and when it comes to chaotic systems, few are more dramatic than the weather. It was from studying weather that the existence of chaos was first realized. And it all started because of our desire to know what is coming up in the future.

3
Weather Worries and Chaotic Butterflies

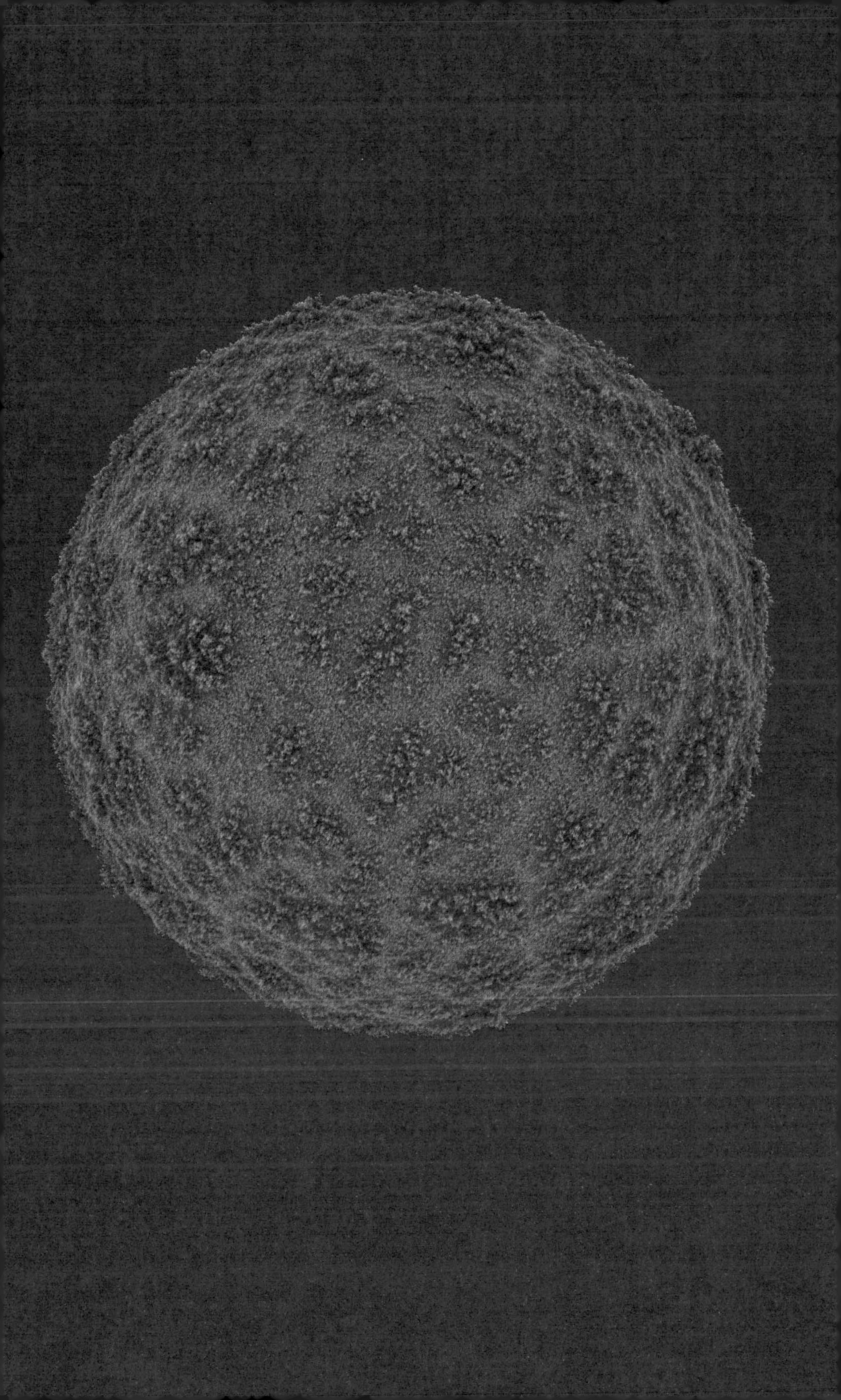

Looking to the future

"Predictions can be difficult,
especially about the future."
Niels Bohr, 1885–1962

What's next?

Perhaps the defining aspect of being human is our desire to know what is on the horizon. What will happen in the future? Not only does this urge permeate every aspect of daily life, it's also the basis for all fiction, from the novel to the movie or TV show. It's the need to find out what comes next, the revelation of unexpected twists and turns, that keeps us reading or watching.

Oracles and prophets, astrologers and soothsayers have provided guidance on the future as long as history has been recorded, usually passing on what was believed to be divinely inspired information. There was no need to make use of numbers and patterns, as the predictions were thought to come from an external, supernatural source. On a more practical level, though, the ability to make realistic short-term local predictions must surely have been responsible at least in part for the early success of *Homo sapiens*.

→
The Oracle at Delphi
King Aigeus consults the Pythia, as the Delphic Oracle was known, in this image on a *kylix* (wine-drinking cup) from around 440 BCE.

Our species dates back around 200,000 years—for much of that time, hunters will have anticipated the future action of their prey to set a trap, acting not as part of some innate response, but by predicting what would happen—*if we chase it this way, it will probably go that way*—and making a successful outcome possible. Similarly, the move from a hunter-gatherer lifestyle to an agrarian society involved a greater requirement to visualize what would happen in the future. For the last 10,000

years or more, increasing numbers of humans have relied on farming for food, an activity that requires awareness of the seasons and how future patterns of weather influence the crops, while raising animals effectively requires an ability to think through the future lifetime of the livestock and plan accordingly.

Some archaeologists believe that the need to make an agricultural forecast means that, for example, our view of the famous stone circle at Stonehenge in Wiltshire, England is literally back to front. The monument is strongly associated with the midsummer sunrise, when the Sun's rays align with some of the stones. Modern-day druids and other sun-worshippers gather at the summer solstice to welcome the Sun's rising toward its zenith. But the midwinter sunrise, attended by far fewer, also presents an alignment—and midwinter, when the days begin to grow longer, was far more important for farming future-gazers than midsummer ever was.

Solstice
The solstices are the days on which the Sun reaches the limit of its daily change of path. The summer solstice is the longest day and the winter the shortest. In the northern hemisphere, these occur on June 20 or 21 and December 21 or 22 respectively, and vice versa in the southern hemisphere.

Broadly speaking, the seasons are relatively easy to forecast. Each year in a particular region there is a familiar pattern to the seasons—there will be some variation from year to year, but broadly it is possible to say how, for example, the weather in winter will compare with the weather in summer or, nearer the equator, the wet season compared with the dry season. But this was by no means the limit of attempts to predict what would happen next. Some individuals have always been prepared to take on randomness itself.

Roll of the dice

In exploring randomness, we have already come across the use of dice and other mechanisms used to support games of chance. It's worth asking why it is that the most obvious examples of randomness tend to come from such games. The fact is that as far back as recorded history allows, humans have been fascinated by the way that bringing a random influence into play will influence the possible future.

In some cases, this was (and remains) a mechanism to support gambling. Here, the ability to predict the future accurately is the hope of every gambler. All too often, a gambler will believe that they have a system that will enable them to "beat the odds." In reality, by definition randomness does not allow for the prediction of specific outcomes. But understanding probability does give the player the advantage of knowing the likeliness of any potential outcome, enabling him or her to act accordingly.

Some randomly driven games have limits imposed by their format that make it easier to predict the future statistically if a player has the appropriate skill. So in a card game like blackjack where cards are drawn from a fixed collection of packs, and all the cards are revealed to the players as the game progresses, it is possible to count the cards as they are played, changing the probabilities that other cards will be drawn as the game progresses.

To use a trivial example, if I am drawing randomly from a single pack of cards and two of the sixes have already been played, but no tens have been played, I know that it's twice as likely that a ten will come up as a six. Casinos use multiple packs shuffled together, so the calculations required and the load on the memory are heavier—but counting is still a helpful way to get a better idea of what is still to come. Casinos consider card counting cheating, though strictly speaking, all the players are doing is playing to their best ability given the information available. It's the casino that is cheating by not accepting counting.

Gambling generates a kind of forecasting mythology, based on a misunderstanding of the nature of probability. Take, for example, a simple repeated coin toss. In one of my talks, I toss a coin with which I have previously thrown nine heads in a row in front of the audience. I ask the audience whether this throw is more likely to be a head, more likely to be a tail or if it has a 50:50 chance, equally likely to be either. Almost always, some of the audience members will say it's more likely to be a tail. This prediction is known as the gamblers's fallacy.

It feels right that after a lot of heads in a row, the next throw is more likely to be a tail—because we know that, on average over a long series of throws, there will be roughly the same number of heads and tails thrown. However, remember the flaw in this reasoning: the coin has no memory. It doesn't know that there have been nine heads in a row. So, with a fair coin, the outcome on the specific tenth throw has still to be 50:50. (In fact, as some audience members realize, by far the easiest way to get nine heads in a row is to use a double-headed coin, so actually the mostly likely next throw is another head.)

It is as a result of such an "It feels right" situation, where common sense deviates from probability, that we are most likely to come up with forecasts of the future that are drastically wrong. The problem arises in gambling and in sports punditry,

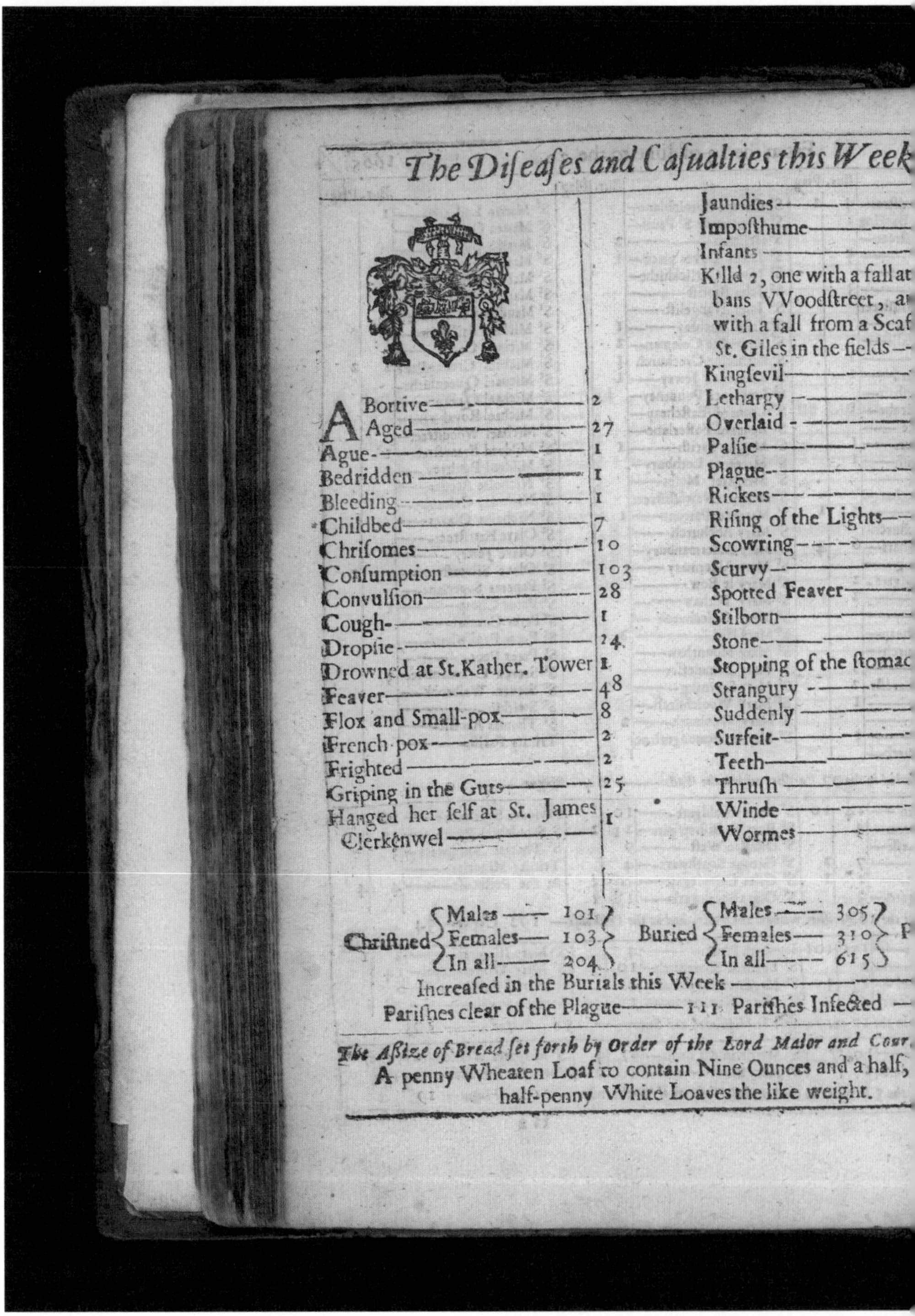

The Diseases and Casualties this Week

Abortive	2
Aged	27
Ague	1
Bedridden	1
Bleeding	1
Childbed	7
Chrisomes	10
Consumption	103
Convulsion	28
Cough	1
Dropsie	24
Drowned at St. Kather. Tower	1
Feaver	48
Flox and Small-pox	8
French pox	2
Frighted	2
Griping in the Guts	25
Hanged her self at St. James Clerkenwel	1

Jaundies
Imposthume
Infants
Killd 2, one with a fall at ... bans VVoodstreet, a... with a fall from a Scaf... St. Giles in the fields
Kingsevil
Lethargy
Overlaid
Palsie
Plague
Rickets
Rising of the Lights
Scowring
Scurvy
Spotted Feaver
Stilborn
Stone
Stopping of the stomac...
Strangury
Suddenly
Surfeit
Teeth
Thrush
Winde
Wormes

Christned	Males	101	Buried	Males	305
	Females	103		Females	310
	In all	204		In all	615

Increased in the Burials this Week

Parishes clear of the Plague 111 Parishes Infected

The Assize of Bread set forth by Order of the Lord Maior and Cour...
A penny Wheaten Loaf to contain Nine Ounces and a half, ...
half-penny White Loaves the like weight.

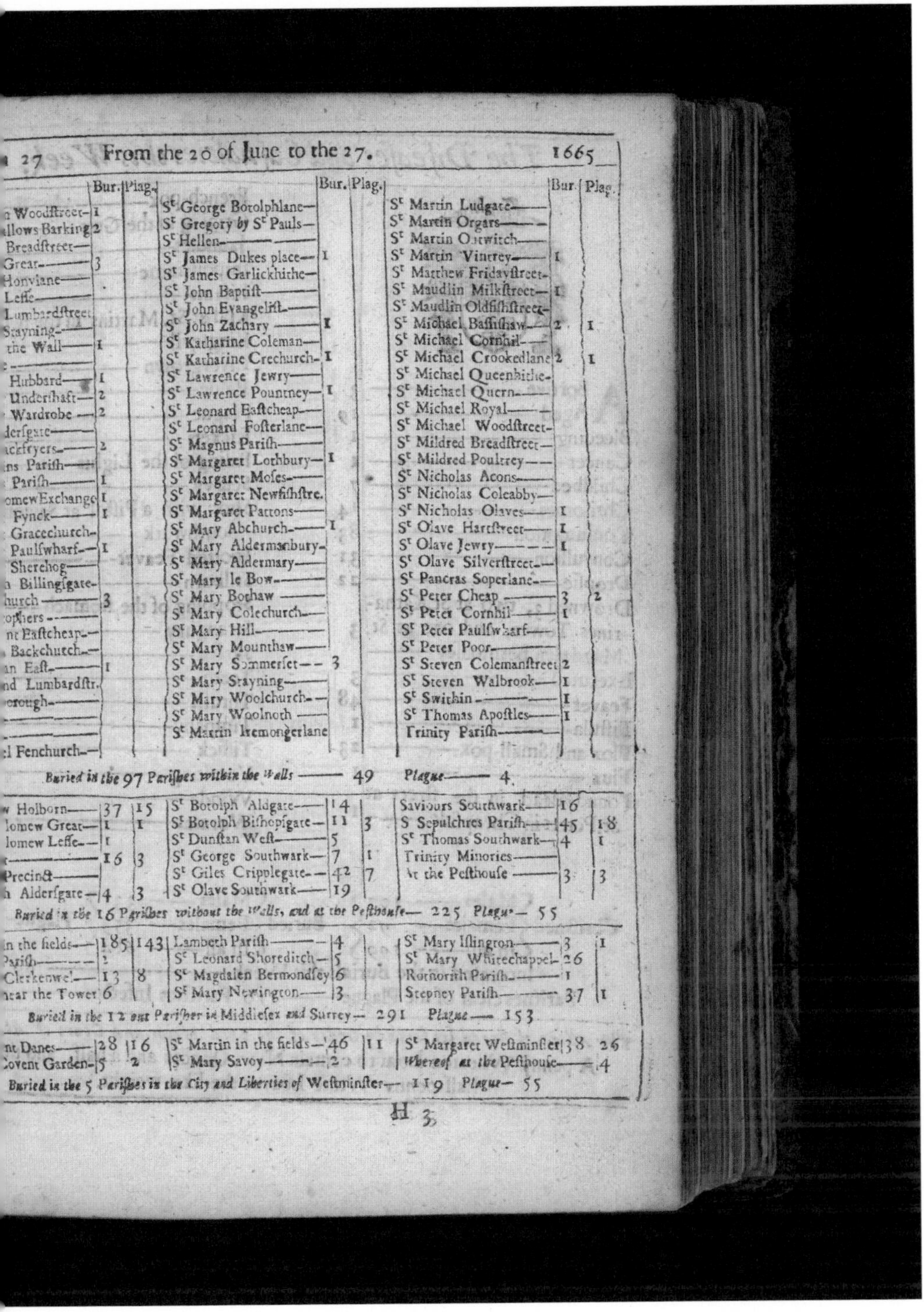

27 From the 20 of June to the 27. 1665

	Bur.	Plag.		Bur.	Plag.		Bur.	Plag.
n Woodstreet—	1		St George Botolphlane—			St Martin Ludgate—		
allows Barking	2		St Gregory by St Pauls—			St Martin Orgars—		
Breadstreet—			St Hellen—			St Martin Outwitch—		
Great—	3		St James Dukes place—	1		St Martin Vintrey—	1	
Honylane—			St James Garlickhithe—			St Matthew Fridaystreet—		
Lesse—			St John Baptist—			St Maudlin Milkstreet—	1	
Lumbardstreet			St John Evangelist—			St Maudlin Oldfishstreet—		
Stayning—			St John Zachary—	1		St Michael Bassishaw—	2	1
the Wall—	1		St Katharine Coleman—			St Michael Cornhil—		
			St Katharine Crechurch—	1		St Michael Crookedlane	2	1
Hubbard—	1		St Lawrence Jewry—			St Michael Queenhithe—		
Undershaft—	2		St Lawrence Pountney—	1		St Michael Quern—		
Wardrobe—	2		St Leonard Eastcheap—			St Michael Royal—		
dersgate—			St Leonard Fosterlane—			St Michael Woodstreet—		
ackfryers—	2		St Magnus Parish—			St Mildred Breadstreet—		
ins Parish—			St Margaret Lothbury—	1		St Mildred Poultrey—		
Parish—	1		St Margaret Moses—			St Nicholas Acons—		
omew Exchange	1		St Margaret Newfishstre.			St Nicholas Coleabby—		
Fynck—	1		St Margaret Pattons—			St Nicholas Olaves—	1	
Gracechurch—			St Mary Abchurch—	1		St Olave Hartstreet—	1	
Paulswharf—	1		St Mary Aldermanbury—			St Olave Jewry—	1	
Sherehog—			St Mary Aldermary—			St Olave Silverstreet—		
n Billingsgate—			St Mary le Bow—			St Pancras Soperlane—		
hurch—	3		St Mary Bothaw—			St Peter Cheap—	3	2
ophers—			St Mary Colechurch—			St Peter Cornhil—	1	
nt Eastcheap—			St Mary Hill—			St Peter Paulswharf—		
Backchurch—			St Mary Mounthaw—			St Peter Poor—		
an East—	1		St Mary Sommerset—	3		St Steven Colemanstreet	2	
nd Lumbardstr.			St Mary Stayning—			St Steven Walbrook—	1	
erough—			St Mary Woolchurch—			St Swithin—	1	
			St Mary Woolnoth—			St Thomas Apostles—	1	
			St Martin Iremongerlane			Trinity Parish—		
el Fenchurch—								

Buried in the 97 Parishes within the Walls — 49 *Plague* — 4.

	Bur.	Plag.		Bur.	Plag.		Bur.	Plag.
w Holborn—	37	15	St Botolph Aldgate—	14		Saviours Southwark—	16	
lomew Great—	1	1	St Botolph Bishopsgate—	11	3	S Sepulchres Parish—	45	18
lomew Lesse—	1		St Dunstan West—	5		St Thomas Southwark—	4	1
—	16	3	St George Southwark—	7	1	Trinity Minories—		
Precinct—			St Giles Cripplegate—	42	7	At the Pesthouse—	3	3
h Aldersgate—	4	3	St Olave Southwark—	19				

Buried in the 16 Parishes without the Walls, and at the Pesthouse — 225 *Plague* — 55

	Bur.	Plag.		Bur.	Plag.		Bur.	Plag.
in the fields—	185	143	Lambeth Parish—	4		St Mary Islington—	3	1
Parish—	2		St Leonard Shoreditch—	5		St Mary Whitechappel—	26	
Clerkenwel—	13	8	St Magdalen Bermondsey	6		Rothorith Parish—	1	
near the Tower	6		St Mary Newington—	3		Stepney Parish—	37	1

Buried in the 12 out Parishes in Middlesex and Surrey — 291 *Plague* — 153

	Bur.	Plag.		Bur.	Plag.		Bur.	Plag.
nt Danes—	28	16	St Martin in the fields—	46	11	St Margaret Westminster	38	26
Covent Garden—	5	2	St Mary Savoy—	2		*Whereof at the Pesthouse*—	4	

Buried in the 5 Parishes in the City and Liberties of Westminster— 119 *Plague* — 55

H 3

London Bills of Mortality
This document from 1664/5, similar to that used by John Graunt (see page 90), shows the causes of deaths in London and burial locations.

but also, for example, in analyses of the stock market. Good forecasters need to have a solid grounding in probability and statistics to avoid falling into the pitfalls that probability presents.

Births and deaths

It was by playing around with statistics that the first useful mathematical forecasts were made by a well-off English button merchant, John Graunt, based in London. Graunt used "bills of mortality"—details of how people had died in London between 1604 and 1661—and birth records in the same time period, to try to understand better what was happening to the population.

As well as analyzing the data he had, Graunt tried to use his statistics to look forward, estimating the distribution of ages at death of a group of people who were born at the same time. This approach was taken up by others around the end of the seventeenth and start of the eighteenth century in London, notably the astronomer Edmond Halley. These individuals, working together in groups based in London coffee houses, started what would become the insurance industry. To make sensible offers for insurance premiums and payouts—effectively bets on the future of individuals—the insurance companies had to make forecasts of how, on average, the lives and deaths of their clients would play out in a complex future.

Coffee houses
Cafés specializing in coffee reached the UK in the 1650s and became popular meeting places for political and philosophical discussion and to undertake business.

Here, for the first time, forecasting moved from being based on guesswork or regularly repeating patterns, such as the seasons, to taking a statistical view across a population, making use of probability and the assumption that (on the whole) randomness was at play to produce a realistic view of some aspects of the future.

But it would be some time later before the most familiar aspect of forecasting took off. It's a forecast many of us check every day: the weather.

Come rain or shine

"Meteorology has ever been an apple of contention, as if the violent commotions of the atmosphere induced a sympathetic effect on the minds of those who have attempted to study them."
Joseph Henry, 1797–1878

Whatever the weather

We can surmise that seasonal forecasting probably dates back at least to Stonehenge, the first parts of which date from around 5,000 years ago. But this only gives a broad overview, taking in months at a time. What farmers, sailors, and anyone else whose lives were affected by the weather really wanted to know was what things would be like tomorrow, or for the next few days. Would it be dry for that long journey you needed to make tomorrow, or would rain spoil the harvest if it wasn't gathered in the next few days? Traditional forecasting, like traditional medicine, relied on a mix of misunderstanding and folk observations that were based on a degree of statistical validity.

One such observation is the old rhyme "Red sky at night, shepherd's delight; red sky in morning, shepherd's warning," where the happy shepherd is looking forward to good weather the next day. Some folk weather predictions have no good basis, but this one makes a degree of sense. Red skies tend to be associated with relatively high air pressure, which is better at trapping the particles in the atmosphere that scatter the red light from the Sun. High pressure in the evening tends to be moving into a region, bringing relatively good weather. By contrast, in the morning the high pressure is more likely to be on the way out, preceding a deterioration in conditions.

Longer-term traditional weather prediction techniques tended to have less of a realistic basis. Many were based on the behavior of animals or trees, on the assumption that being more attuned to natural cycles, they reflected something we couldn't see—so, for example, if trees bore more berries during fall, it was supposed to presage a harsher winter. But there is no evidence to back up such faulty logic. The trees have no way of telling what will happen a few months down the line. The same goes for the more modern U.S. tradition of Groundhog Day, when the groundhog's reaction to its shadow (or lack of it) is said to foretell the weather for the following six weeks. In reality, the groundhog's predictions are no better than tossing a coin and leaving the forecast to random chance.

Barometers, first invented in the seventeenth century, had become popular in homes by the late nineteenth century, by which time it was realized more explicitly that a shift in atmospheric pressure could be an indication of changing weather patterns. It was around this time also that the weather forecast as an early form of the present concept came into being, made possible by a combination of data collected from widespread weather stations and the electric telegraph, which allowed that data to be pulled together at a central location to attempt a wide-scale forecast for a region or country.

Formal forecasting was first attempted by Francis Beaufort and Robert FitzRoy (the latter better remembered as captain of HMS *Beagle* during Charles Darwin's famous voyage) for the British navy during the 1850s. The first public forecast was published in the London *Times* newspaper in 1861, with the first published weather chart following in 1875. For several decades the forecasting provided was qualitative, based on observations combined with predictions from atmospheric pressure, rainfall, wind directions and strength, but by the 1920s meteorology was a numerical science, breaking up the three-dimensional space occupied by the atmosphere into small regions, with predictions made on how the weather in each would progress over time.

→
Early weather station
An 1880 weather station with an automatic anemometer, recording wind speed and direction.

Despite investments of huge amounts of time, effort, and money into weather forecasting, the outcomes were often wrong, and it took another four decades before anyone realized why, beyond the fact that the weather was complicated. In the 1960s an early computerized weather simulation would reveal the true nature of chaos.

Diverging weather patterns
Lorenz's 1961 printout shows how the weather patterns for heat exchange and atmospheric flows predicted by the two computer runs become more and more divergent.

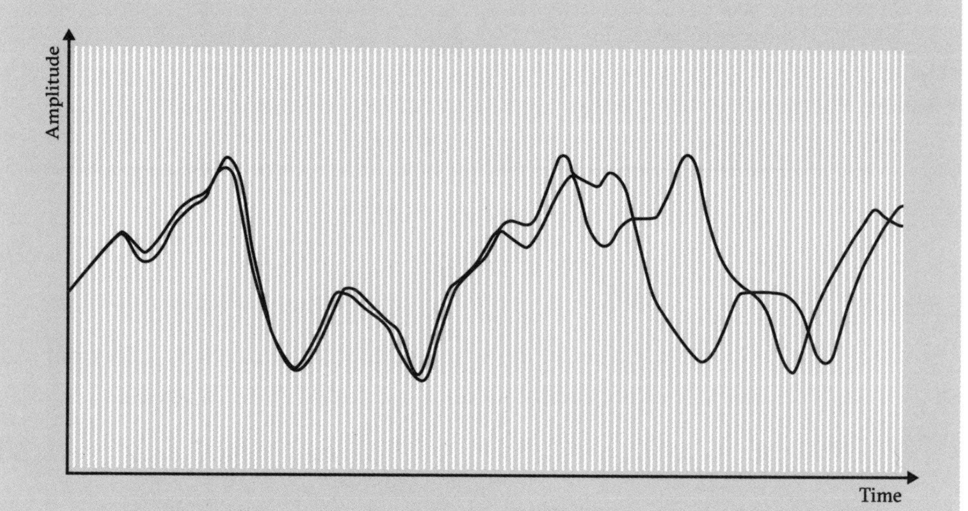

Cutting corners

Simulation was one of the earliest applications of computers, using equations to produce a sequence of numbers that evolved over time, attempting to reflect some aspect of the real world, whether it was the explosion of a nuclear bomb or the patterns of the weather. In 1961, an American meteorologist based at the Massachusetts Institute of Technology, Edward Lorenz, was using a very basic small computer (about the size of a washing machine) on a weather model that made use of limited information such as temperature and wind speed.

As the computer was slow, and the simulation involved a great many calculations taking a long time to run, when Lorenz wanted to extend the study of a particular weather pattern, rather than start over from the beginning, he began the program partway through, using a printout of the values that the model required, taken at that point in the previous run. When he continued the simulation, the output it generated deviated from his original run—not just by a small degree, but producing a totally different forecast.

After checking that the computer was running correctly, the cause of the deviation was tracked down to the numbers Lorenz had input from the printout. When handling real numbers with decimal places, a computer has to make compromises. There are limits on the number of decimal places any computer can handle—and back in the 1960s, those limits were strict. The machine that Lorenz was using worked to six decimal places. So, a number might be represented in its calculations as 1.385262—but it had been programmed to round these numbers to three decimal places when printing the number out. This made sense, as the precision was spurious when compared with reality; real meteorological data was not available to this level of detail. That same number would have been printed out as 1.385.

Decimal places
The number of values appearing after the decimal point. So, for example 3.142 is pi to 3 decimal places, while 3.14157 is pi to 5 decimal places.

This may seem negligent, but scientists are always wary of unjustified precision. There is no point stating a number to many decimal places if your instruments simply don't work to that degree of accuracy in the actual experiment. If a number is specified to more decimal places than can be measured, it seems to imply that more is known about it than truly is. This doesn't mean that the computer was wrong to work to six decimal places—in a calculation, the more decimal places the better—but it was eminently reasonable to print only three.

After repeated runs, it became clear that tiny changes in the values used to set off the simulation resulted in very significant deviations down the line. This was, of course, not a real weather system—it was far less complex—yet even in this simplified example it was clear that something surprising was happening. Lorenz had discovered the principle of mathematical chaos. Although earlier examples, such as the three-body problem (see page 59), had shown the difficulty of making predictions when systems had several interacting components, chaos theory formalized these observations as a mathematical reality that was all too common in the real world. And this would have very significant implications for the business of weather forecasting.

Weather systems

The weather is primarily dependent on the interplay of two huge fluid systems—Earth's atmosphere and the oceans—along with the solid landmass. Weather is traditionally defined in terms of the atmosphere, and the atmosphere lies at the heart of weather modeling, but without bringing in oceans and land masses (including the impact of the built environment), there is not much hope of an effective forecast.

We can see this in the impact of just one of the many seaborne contributions to weather, the Gulf Stream. The presence of this phenomenon means that the northwest of Europe is a massive 16°F (9°C) warmer than it otherwise would be—the climate here should be more like that of Siberia. The name Gulf Stream makes it sound like a flowing river, but in fact the weather phenomenon is more like the loop of a conveyor belt. Winds blowing over the North Atlantic cool the already frigid water, which sinks and flows at a low level toward the equator. At the same time, water on the surface of the sea in the Gulf of Mexico is being warmed by the Sun, and this warm water streams north to compensate for the cold water flowing back, far below it. This process, known technically as thermohaline circulation, transports large amounts of heat in the water from the tropics to northern latitudes. It was the collapse of this "North Atlantic conveyor" that was portrayed so dramatically (and with wild inaccuracies) in the 2004 movie *The Day After Tomorrow*. The scene as portrayed was not realistic because everything happened far too quickly, and because there is no sign that the conveyor is going to stop entirely, but the underlying concept is not entirely fictional.

Thermohaline
A term in oceanography that refers to the temperature and quantity of salt in seawater.

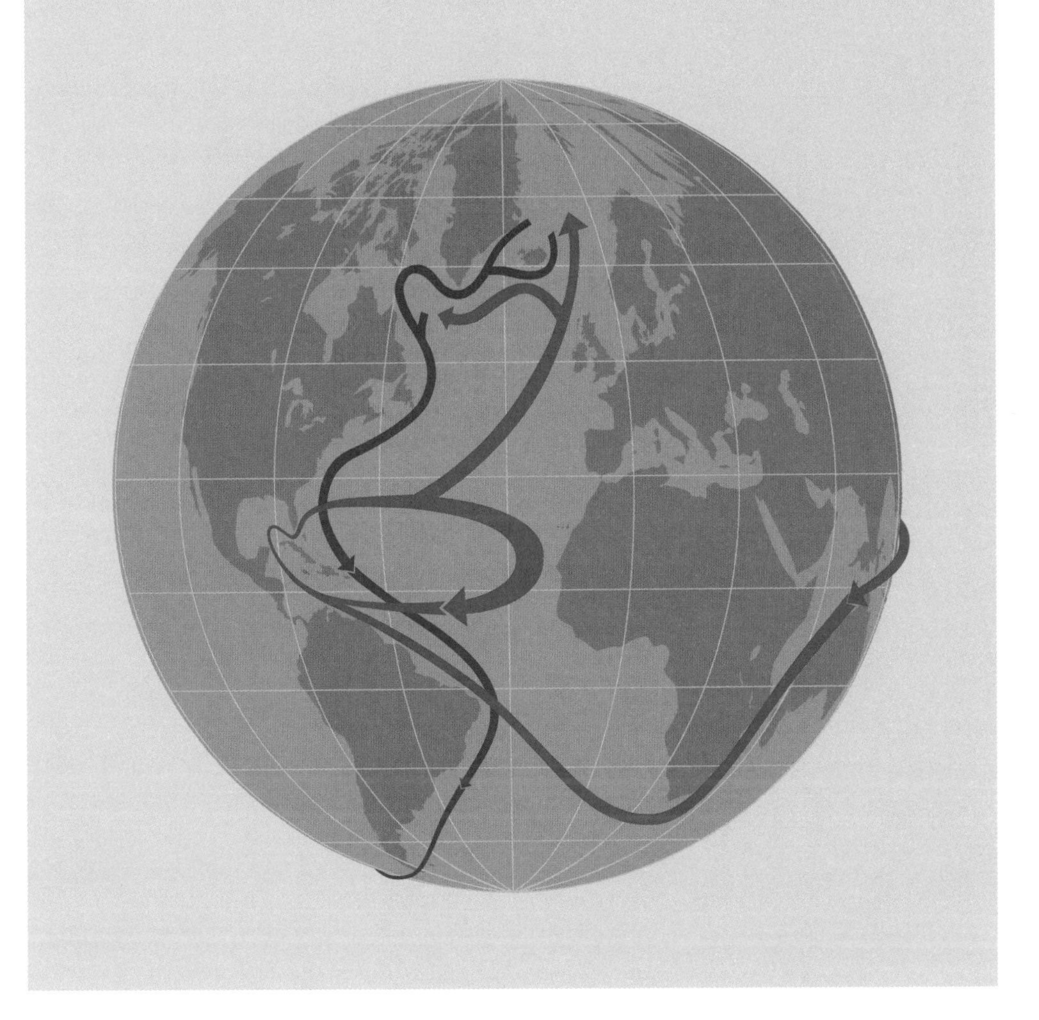

North Atlantic conveyor
The warm Gulf Stream is shown in red and the returning, colder, low-level North Atlantic Deep Water stream in blue.

There is some evidence that a side effect of climate change could slow down the conveyor. As more and more fresh water from melting ice sheets enters the oceans it decreases the density of the cold water that should be diving down and heading south (fresh water is less dense than salt water). The result is to reduce the driving force of the conveyor. So, in this sense at least, *The Day After Tomorrow* is not pure fantasy.

Unlike the event in the movie, though, this slowdown is likely to be a very gradual process—taking perhaps 100 years to reduce the strength of the Gulf Stream by around 25 percent—a change that will be more than balanced out in its impact by predicted global warming. It could even be a good thing for northwest Europe: such a change would mean that this region suffers considerably less from the impact of global warming than other parts of the world.

The Gulf Stream is just one of the contributors to the complexity of weather. As the different components interact, the result makes the three-body problem of gravitation look totally trivial. We clearly can't work at the level of individual atoms in the atmosphere, but the sheer complexity of weather systems makes it obvious why the atmosphere was the natural place for chaos theory to be first applied. Whether or not we agree with a weather forecast, though, the weather has generated far less argument that a related essential: predicting the future climate.

The big picture

"Does the flap of a butterfly's wings in Brazil
set off a tornado in Texas?"
Edward Lorenz, 1917–2008

Climate concerns
Weather is what we experience directly, day to day, but climate gives us the bigger picture, the overall pattern of weather across time and space. As Mark Twain put it, "Climate is what we expect, weather is what we get." The difference between climate and weather is why so many people misunderstand climate change. They experience miserable weather or a particularly cold winter and conclude that global warming obviously doesn't exist.

The reality is that the climate as a whole is undergoing change, a change that includes warming when averaged across the planet, but at any one point in space and time we are not going to experience that average. Because our personal experience is always of the weather, not of the climate, it is easy to underestimate what the threats of global warming really mean. A temperature rise of a handful of degrees sounds trivial. After all, temperatures vary by tens of degrees between summer and winter. The predictions of global warming don't sound dramatic—but that is because they represent an average, reflecting far greater extremes of heat and cold at any particular place and time.

Climate change is nothing new. In fact, it's perfectly natural. Over the 4.5 billion years that Earth has existed, the climate has always been in flux. Go back to the period where the

Ice Age Earth
Approximate extent of the ice sheet during the most recent maximum extent of the current ice age.

earliest-known signs of life have been spotted, around 3.7 billion years ago, and conditions were hot—significantly more so than the current direst predictions for climate change—with temperatures around 18°F (10°C) higher than they are now.

Between then and the present, conditions have also been much colder. There have been at least four major ice ages in the lifetime of the planet. These are periods when the polar ice sheets extended much farther onto the continents than they do today, transforming the environment in a drastic way. Apart from the damage caused by the sheer weight and thickness of the ice and the difficulty for life to continue existing in higher latitudes as the ice sheets encroached, global temperatures plummeted 18°F (10°C) below the current levels.

Pliocene-Quaternary
"Quaternary" refers to our current geological period, stretching back around 2.5 million years, while "Pliocene" refers to the epoch (a subdivision of a period) immediately before the Quaternary.

Technically, we are still in an ice age, called the Pliocene-Quaternary, a phase in Earth's climate that began around 2.5 million years ago. As an ice age progresses, there are periods known as glacials, when the ice is advancing, and interglacials when the ice retreats—we have been experiencing an interglacial for around the last 11,000 years. During this time (and it's no coincidence that this is the time frame during which all the great human civilizations have arisen), life on Earth has been relatively easy.

If there were no human interference, the interglacial would be expected to end somewhere between 1,000 and 25,000 years from now, plunging us back into the grip of the ice. Much of Canada, the northern United States, and northern Europe would be hidden under sheets of ice once more. One potential benefit of global warming is that human activities have almost definitely already done enough to stop the next glacial period occurring. (Among all the justifiably alarming headlines we have to remember that warming is not inherently bad.)

The human impact

You will sometimes hear those concerned about the changing climate and the impact of humans on it say that we need to "save the planet." The planet, the ball of rock, is fine. There is nothing we can throw at Earth that it can't get over in a million years or two, a tiny fraction of its lifespan. What we want to do is not to save the planet but the life upon it and particularly to save our civilization, or at the very least to preserve human existence, which is much more fragile and at the whim of the climate than is the future of the planet itself.

Some kind of action is definitely necessary to do this, as there is good evidence, which has persuaded the vast majority of scientists around the globe, that human activity over the last 150 years is making a significant contribution to climate change. Primarily through increases in the levels of greenhouse gasses from farming, industry, homes, and transport, we are indubitably impacting on the climate. And since the 1960s, the rate of global warming has increased. Between 1906 and 2005, global average temperatures rose by 1.26°F (0.74°C), and the rise was significantly greater during the second half of the century than the first.

You can see the impact of climate change in what climate scientists rather confusingly call global temperature anomalies. These reflect the way temperature is measured. It is all very well to ask how the average temperature on the entire planet is varying—but how do you find out the average temperature of such a huge body as Earth, with such varied weather conditions at any one time? It isn't actually possible to calculate a meaningful average for the whole world. Not least, there isn't a good enough spread of weather stations, evenly across Earth's surface, to record the measurements required to achieve this.

Instead, climate scientists use a measurement that isn't an absolute value, but a relative variation, which they call a temperature anomaly (which might be interpreted as a strange temperature, but "anomaly" is used here to mean a variation.) In finding the anomalies, climate scientists compare average temperatures for any particular period against the long-term averages produced by the same weather stations. This way, any variation (the anomaly) in the average temperature sticks out from the rest of the data.

This is the main reason there is some variation in the average temperature comparisons given by different climate monitoring bodies around the world, because the size of the anomaly depends on the long-term period used for the comparison and some organizations have decided to use different periods. This means that while they won't necessarily agree on, say, which year was the hottest on record—yet all concur that the decade from 2010 onward was the hottest since records began, as was the decade before it.

Far more questionable is the accuracy of models predicting what impact this warming will have on the environment, as the

Global temperature anomalies
Difference in temperature from the 1951–1980 average in degrees Celsius. The black line is a running 5-year average to smooth out the change.

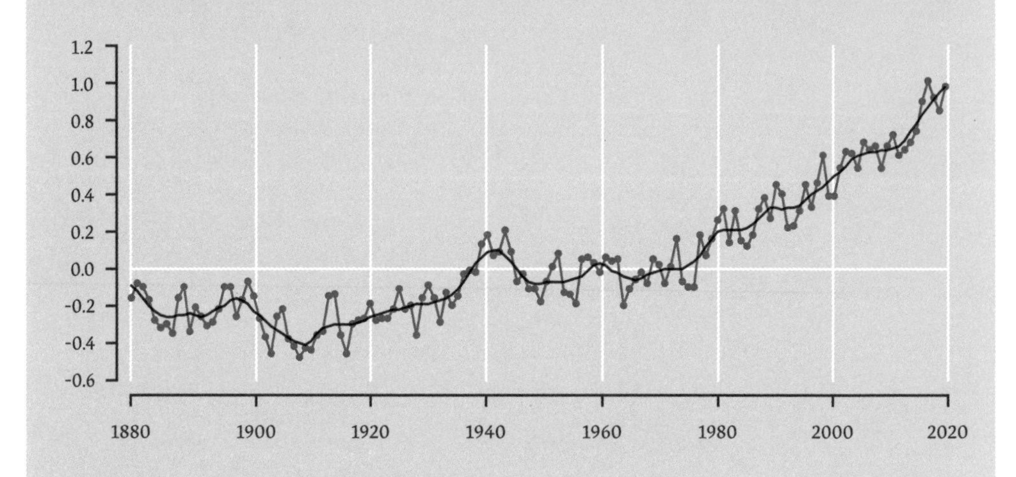

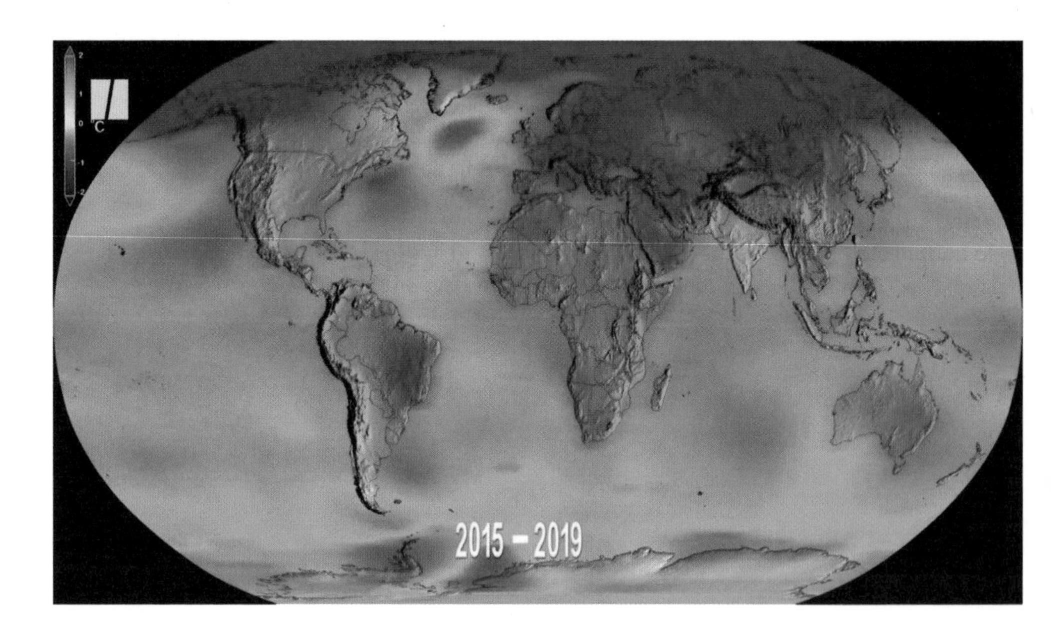

number of variables is huge, from the complex impact of cloud cover to the melting of ice sheets. The effect that clouds have is particularly difficult to model because, depending on their altitude, clouds can either increase or decrease the level of global warming (high clouds trap heat radiation from Earth, preventing its escape; low clouds, being thicker, reflect more solar energy before it reaches Earth), and there is no sensible way to build their contribution into a computer model.

That said, the best models are quite effective at "predicting" the past—when fed with earlier data, they match what has happened over the last 100 years reasonably well—and hence have a fair chance of predicting the future. And there is no doubt that warming with a human component is occurring. The only disagreement between the models is how fast our contribution is heating things up. Like weather models, climate models make use of multiple runs to see how sensitive they are to variables (see "Ensemble solutions," page 109), but there is less of an issue with the chaotic system here as the climate models are taking a step back to get the big picture and need not concern themselves with the detailed interactions of the components that shape the weather.

Climate forecasts suggest to us that relatively small changes in global temperature can have huge impact on day-to-day lives—but the amplification of outcome is trivial when compared with Edward Lorenz's other main contribution to the study of chaos: the butterfly effect.

The butterfly effect

Most of us are familiar with a variant of the title of Lorenz's original talk that would give this concept a name: "Does the flap of a butterfly's wings in Brazil set off a tornado in Texas?" The idea is simple. Chaos theory tells us that a small change in initial conditions will result in a large change in outcome. Could that change be as tiny as the flap of an insect's wings, and could it result in as immense an outcome as a tornado, causing devastation thousands of miles away?

←
Global temperature anomalies 2015–2019
Compares average land and sea temperatures to the 1951–80 average. Areas that are hotter than normal are red and cooler areas are blue.

What tends to be forgotten is that Lorenz's answer to the question was "No." There is small and there is ridiculously small. What is small in terms of a weather system is still quite large compared with the airflow generated by a flap of a butterfly's wings. Such very minor inputs do tend to be damped out in a complicated system, dying away rather than being magnified. For that matter, a tornado is a localized weather pattern with limited influence from any distance. Yet Lorenz's underlying concept of small differences having huge outcomes was not wrong.

Despite the exaggeration, "the butterfly effect" remains a very effective reminder of the central message of chaos—that a small difference initially can lead to big changes down the line. It's something we should really be aware of, as it's a common occurrence in everyday life. The 1998 movie *Sliding Doors* provides a good illustration of this, where the central character has two very different potential futures, depending on whether or not she catches a particular train on the London Underground. We all have very small events in our lives that have had a huge influence on later outcomes—in this respect at least, it's not unreasonable to describe your life as being chaotic.

The title of the talk, which Lorenz would later claim was made up by a colleague when he himself failed to come up with one in time, was a provocative reminder of this, if not meant to be taken literally. It's also worth noting that, though weather is a chaotic system, the original issue that Lorenz studied was not chaos in the weather itself but in a computer model of it. Sometimes, the issue with weather forecasting difficulties can be caused by chaotic forecasting that does not necessarily reflect specific chaos in the weather system itself. Luckily, there is a way to deal with forecasting chaos that can help deal with both types of contribution.

Transforming meteorology

"People never pay attention to weather reports; this, I believe, is a constant factor in man's psychological makeup, stemming probably from an ancient distrust of the shaman. You want them to be wrong. If they're right, then somehow they're superior, and this is even more uncomfortable than getting wet."
Roger Zelazny, 1937–1995

Forget the long term
Ever since weather forecasts were first published, our relationship with them has been fractious. Historically, they were all too often wrong. Infamously in the U.K., a TV meteorologist called Michael Fish gave a reassuring forecast on the evening of October 15, 1987. He commented "Earlier on today, apparently, a woman rang the BBC and said she had heard there was a hurricane on the way. Well, if you're watching, don't worry, there isn't." A few hours later, much of England was hit by the worst storm it had experienced in three centuries, resulting in the deaths of 18 people, the toppling of an estimated 15 million trees, and widespread damage to vehicles, property, and infrastructure as well as wildlife.

The error in Fish's forecast was an extreme case, but weather forecasts at the time were often the subject of jokes for their inaccuracy. Forecasts up to five days ahead have improved immensely since the 1980s, but what is bizarre is that, despite

Lorenz's discovery, perhaps in response to our desire to know the future, long-term forecasts continue to be published to this day. From Lorenz's work it became clear that it would *never* be possible to sensibly forecast weather more than ten days ahead. Next time you hear several months in advance that there is going to be a particularly hot summer, or heavy snow in winter, bear in mind that simply reporting what conditions are usually like at a location at a particular time of year is significantly more accurate than attempting a long-term forecast based on a chaotic system.

Once Lorenz had understood the impact of chaos he immediately realized how farcical long-range forecasts were. He commented:

"We certainly hadn't been successful doing that anyway and now we had an excuse. I think one of the reasons people thought it would be possible to forecast so far ahead is that there are real physical phenomena for which one can do an excellent job of forecasting, such as eclipses, where the dynamic of the sun, moon and earth are fairly complicated, and such as oceanic tides.... Tides are actually just as complicated as the atmosphere. Both have periodic components—you can predict that next summer will be warmer than this winter. But with weather we take the attitude that we knew *that* already. With tides, it's the predictable part we're interested in, and the unpredictable part is small, unless there's a storm."

Ensemble solutions

A huge change happened in the way computerized weather forecasting was undertaken toward the end of the twentieth century, which would allow for significantly greater accuracy in forecasts covering the next 24 hours to five days. The earlier computer forecasts had worked on a single picture of how things were going to be. Forecasters would run the model, and the output, describing the future evolution of the weather system, would be produced. The problem is that, as we now know, weather systems are hugely dependent on the exact initial conditions, so any particular forecast was highly likely to be wrong.

Now, with vastly more computer power available to them, meteorologists run a model many times, each with subtly different changes, reflecting the uncertainties in the data and how the weather will evolve. The European Centre for

←

Low tide, Oregon
Unlike the weather, the periodic behavior of the tidal system happens on a twice daily basis, making it more usefully predictive.

Medium-Range Weather Forecasts in Reading, England, for example, generally regarded as the best in the world, supplies such "ensemble" forecasts internationally, typically running 51 forecasts per day on a collection of supercomputers, each varying slightly in its parameters. The outcomes are grouped together by those with similar results for an indication of the most likely forecast.

Supercomputer
A large, specialist computer, typically using thousands of processors to enable extremely rapid calculations widely used in weather forecasting.

This ensemble approach means that it is possible to get a much better picture of the probability of different weather events occurring, and it is why probabilities ("a 40 percent chance of rain between 11:00 and midday," for example) are now much more likely to be shown on weather forecasts. Probabilities, though, can cause some confusion: that statement does not mean there will be rain in 40 percent of the places covered, or that there will be rain for 40 percent of the time period to which it applies. It means that 40 percent of the model runs forecast rain at some time in the period for this region. For some reason, percentage-based forecasts became popular earlier in the U.S. than the U.K., where forecasters thought that the public was not comfortable with probability and needed a firm statement on the upcoming weather. Now, though, such forecasts have is still commonplace.

As a result of better models and ensemble forecasting, a quiet revolution has occurred. Just 30 years ago, forecasts were more often wrong than right. Today, short-term forecasts are much more reliable. We have become used to this change and still grumble when the forecaster gets it wrong—but we have reason to complain far less often than used to be the case before ensemble forecasts were introduced. Thanks to a combination of excellent satellite observation and modern forecasting methods and technology, forecasts can usually be accurate over 24 hours and reasonably effective over three to five days, but anything longer than this become little better than educated guesswork.

Long-range forecasting remains much more of a dark art than science, because the potential for change is so great over a longer period of time. It is certainly true that many long-range forecasts, no matter how much computing power is used, remain poor. Most of us have eagerly anticipated the promise of a great summer, only to find that the weather turns out to be a flop. It almost seems more likely for long-range forecasts to be wrong than right.

→
Hurricane Katrina track forecast
Ensemble forecast showing the probability of different tracks for hurricane Katrina in August 2005.

In part, this failure to predict with any accuracy is because a long-range forecast can only give the broad picture, and locally weather can be very different from a national average. But as we have seen, it is also impossible to be correct when taking a chaotic system like the weather so far into the future. Bearing in mind that "what conditions are like at this time of year" is the best forecast we can achieve, things have improved in our understanding of prevalent conditions in the last two decades, as we now have a better appreciation of large-scale, long-lasting weather patterns such as El Niño, part of a long-term weather cycle involving a band of warm water in the central Pacific which, unusually, moves eastward toward South America and can influence the weather in a large region for an entire season.

Weather forecasting as a discipline brought into being a new mathematical study in the form of chaos. Over time, exploring the nature of chaos would uncover whole new oddities in this fascinating field—beginning with circumstances when chaos, oddly, spawns calm.

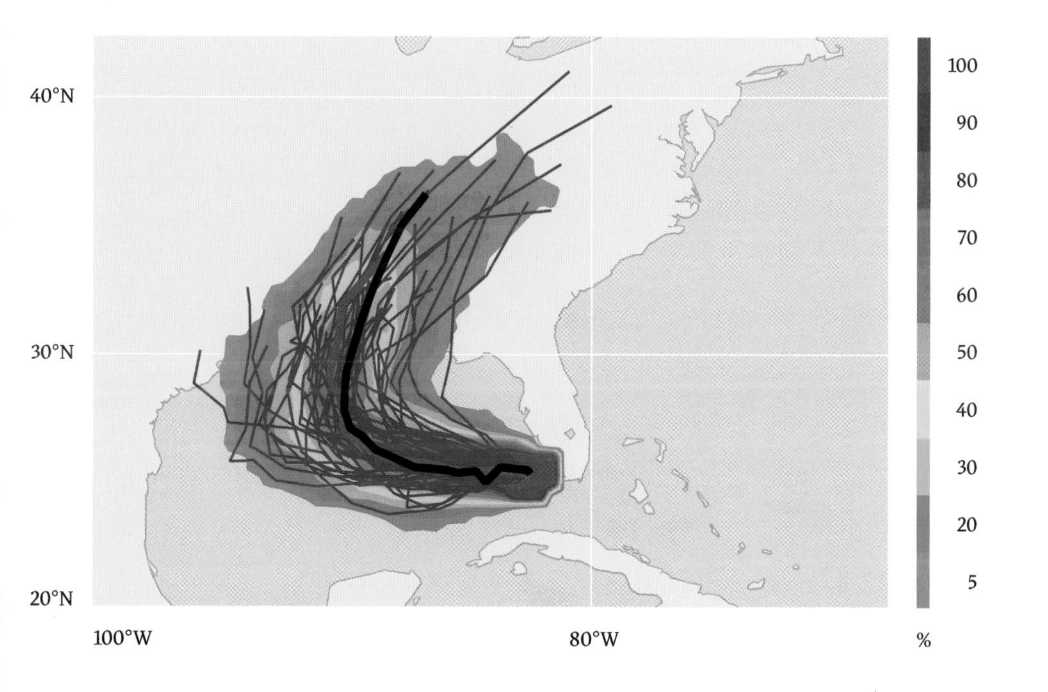

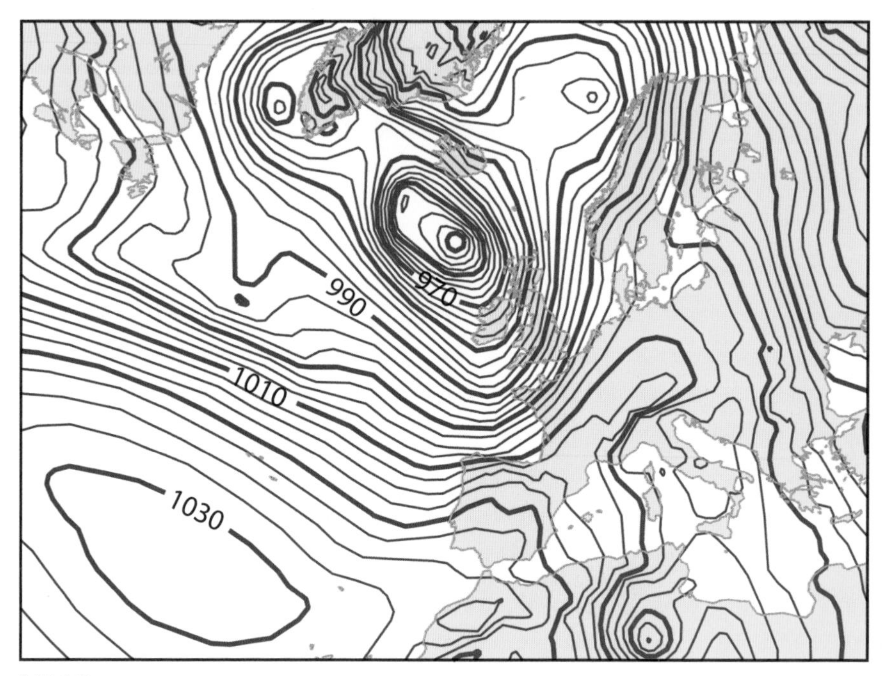

Initial time

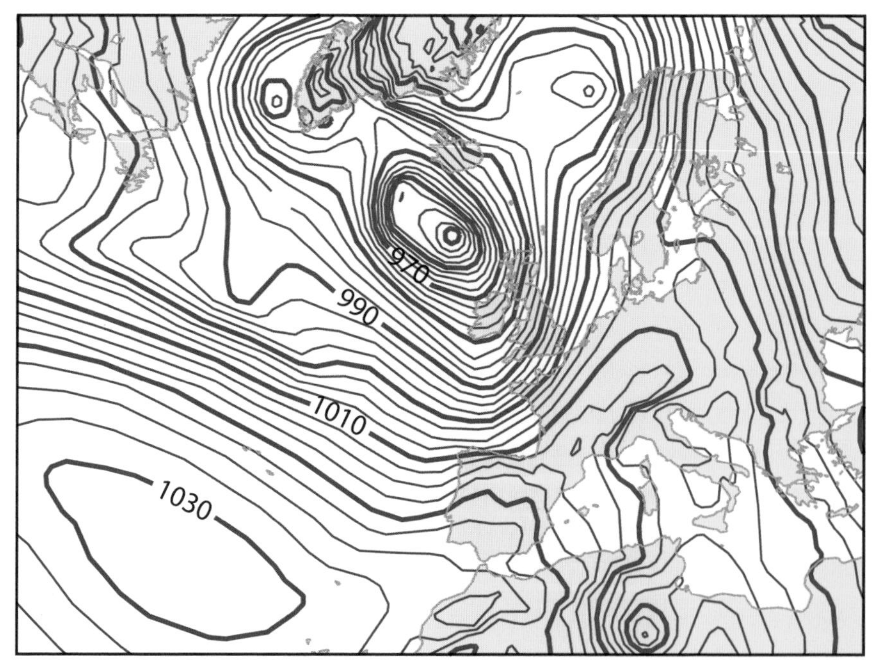

Initial time

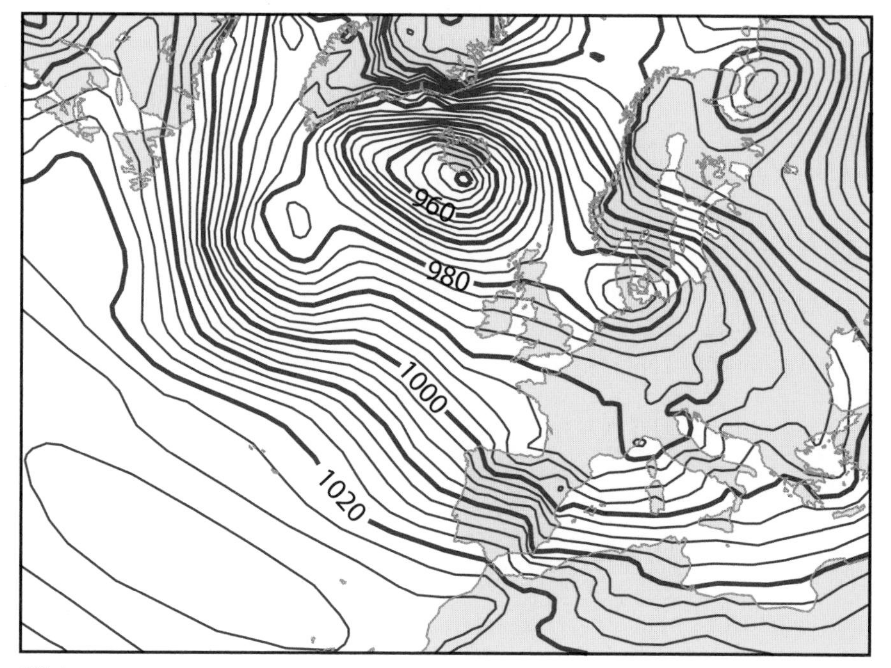

48h forecast

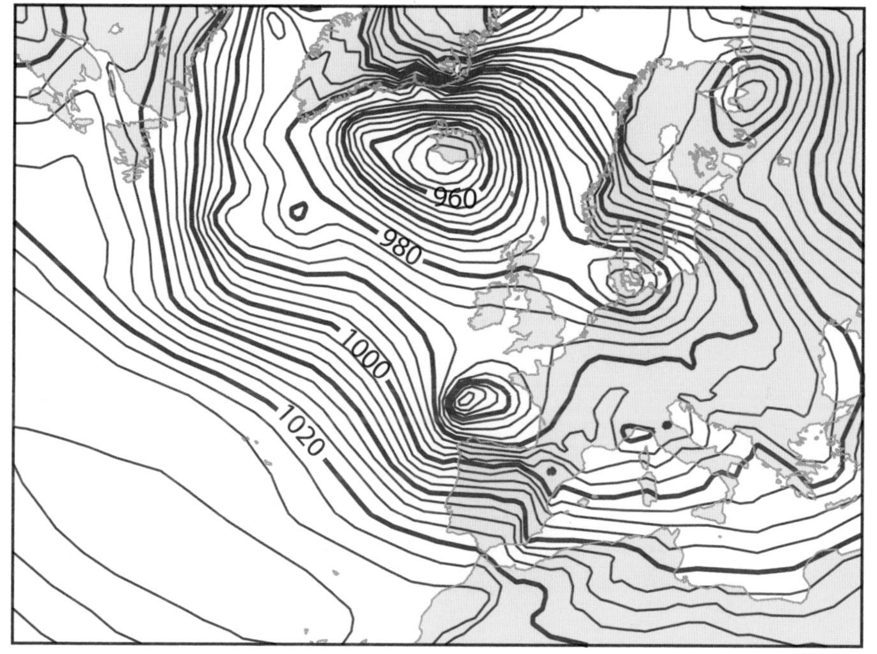

48h forecast

ECMWF ensemble forecasts
Ensemble forecasts made by the European Centre for Medium-Range Weather Forecasting use 50 subtly different starting conditions. Here, two of these for atmospheric pressure show how starting with near identical conditions results in very different 48-hour forecasts.

4
Strange Attractors and Immeasurable Distances

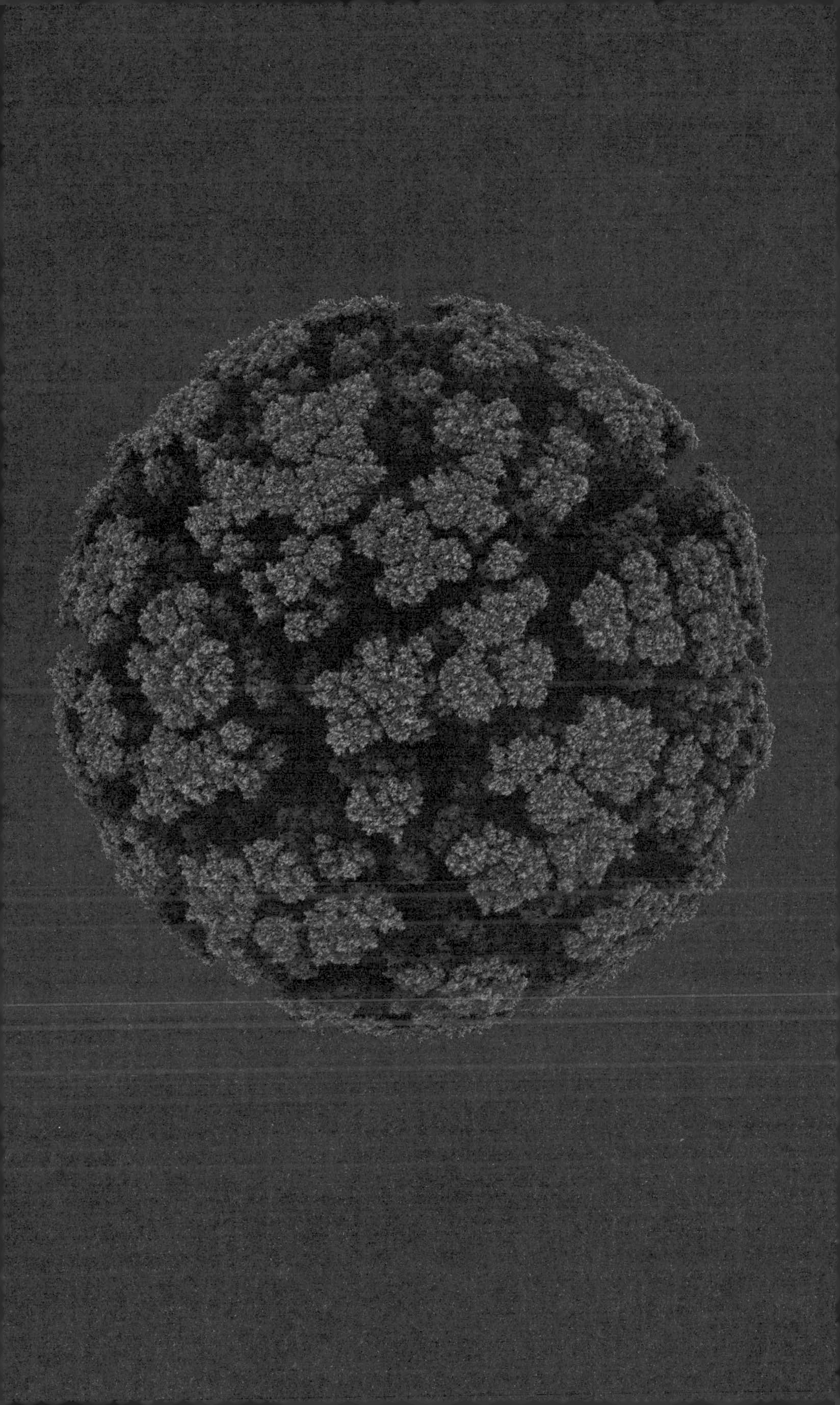

Islands and attractions

"Art has something to do with the achievement of stillness in the midst of chaos."
Saul Bellow, 1915–2005

Calm in chaos
Looking at the way that a chaotic system seems rapidly to run out of control, it would be easy to assume that once chaos has set in, the wild behavior will continue indefinitely. However, real chaotic systems are capable of producing order from within the seeming anarchy. Perhaps the most obvious occurrence is again in the weather where, for example, strong winds can suddenly drop to nothing.

A simple chaotic system where this occurs is one known as a "period-halving bifurcation." Not the catchiest of terms, this is where in effect there is a series of opportunities for a system to go one of two ways, but the time for each choice reduces, resulting in an initially chaotic system that eventually settles down to a stable, calm progress. Confusingly, this is more often referred to as a "period-doubling system," as the number of events before the system repeats itself doubles, even though the time between the doubling reduces each time. Although the system is chaotic, as the time between events becomes extremely small, the result is effectively continuous, and we find a state that is impossible to distinguish from smooth, steady progress.

→
Red blood cells
Seen through a scanning electron microscope, these cells last only a few months, and are constantly replaced in a process that features period-doubling.

A familiar example of such a system is the drip from a faucet. With a very slight drip, water will typically drop regularly. With

slightly more water delivered, the faucet will start to make two distinctive types of drip: drip 1, drip 2, drip 1, drip 2.... A little more water and each of those types of drip itself split into two, so we get: drip 1, drip 2, drip 3, drip 4, drip 1.... And so on, before long coming so frequently that the result is continuous flow. All sorts of systems undergo this kind of behavior, for example blood cell production and predator–prey interactions in the natural world, or the interaction of electronic oscillators in physics. These systems can be described by different but relatively simple equations.

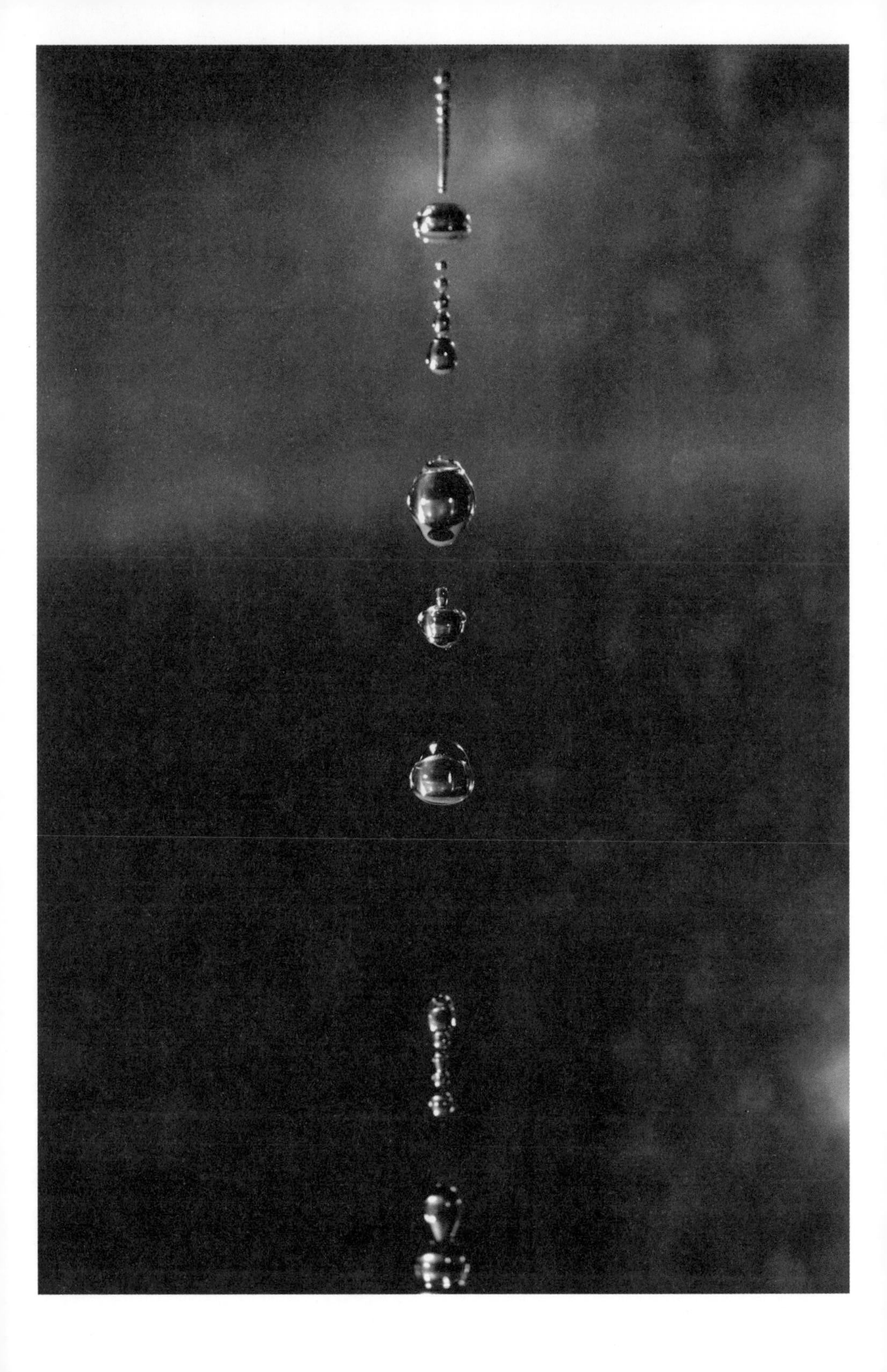

Remarkably, as discovered by American physicist Mitchell Feigenbaum, the equation that describes the system does not matter; all such period-doubling systems have what appears to be a universal constant at their heart. Each time the number of possible modes splits, the factor that is driving the split (temperature, water pressure, or other variable) has to change by a smaller amount than was the case the previous time. Each is 1/4.669 times the size of the previous one. Nature has a handful of universal constants such as pi—and this "Feigenbaum number" of approximately 0.1242 appears to be another.

Chaos in calm

A similar but apparently inverted effect can emerge when some non-chaotic systems are put under pressure. These are systems that we might expect to behave in a predictably calm fashion, but a sudden major change to their environment can cause them to jump into chaos. The work that was first given the name "chaos" was not Lorenz's weather discovery (which Lorenz referred to by the less-catchy title "deterministic nonperiodic flow"), but Australian biologist Robert May's later study of animal populations (see chapter 6).

After May made a surprise discovery that populations growing above a certain rate would exhibit chaotic increases and decreases, he began to look for other possibilities in the natural world where chaos could have an impact. One big surprise came in the field of disease outbreaks. It had been observed that these could have wild swings in the numbers of people infected: here was a chance to apply his new understanding of the mathematics of chaos.

May discovered that, just as a population that is increasing rapidly can go into a chaotic state, the same applies to the spread of a disease when it is hit by a sudden, strong change. And this includes the impact of vaccination. Where you might expect that after the introduction of a major inoculation program, cases of a disease would rapidly drop off, fading down to a steady low level, the mathematics made it clear that the incidence levels were likely to pass through a chaotic phase first, on the way to those low levels.

This meant that, counterintuitively, the number of infections could actually suddenly jump upward after the start of an inoculation program. Without the awareness provided by chaos theory, practitioners might assume that the inoculation

←
Dripping tap
Rather than falling steadily, the drips are period-doubling.

was not working or, even worse, that the vaccine could be infecting some of the patients. An understanding of chaos mathematics did not make it any more possible to predict what would happen from day to day—but it did mean that such apparently anomalous behavior could be understood, making it less likely that a successful program would be interrupted too early before its outcome had settled down.

Uncanny attraction

Although chaos can seem to involve a total lack of organization, in practice chaotic systems usually exhibit inner structure, or operate in such a way that after initial wild chaotic behavior, a broad range of starting points can end up homing in on either the same or a similar outcome—this phenomenon was given the name of an "attractor."

A physical example of an attractor is a flat table with a dip in it. If small balls are rolled across the table relatively slowly, they could end up at any point on the table when friction brings them to a stop. However, any ball that passes over the lip of the depressed section—heading in any direction—will slide into the dip and these balls are likely to end up clustered at the bottom of it. The dip is acting as an attractor. Here the force of attraction is gravitational, though other forces could be deployed—for example, a magnet attracting steel ball bearings—or, in a system of interacting components, it can be simply the dynamics of the system that lead to the attractor forming.

One particular type of attractor to emerge from chaos theory is the mysteriously named strange attractor. This is an attractor that has some parts similar to others—it is fractal (more on fractals in due course) in form. The term "strange attractor" was first used in a paper on turbulence by Belgian physicist David Ruelle and Dutch mathematician Floris Takens.

Fractal
A mathematically derived shape in which small parts are visually similar to the larger whole. Common in nature, for example in plants, clouds, and coastlines.

The strange world of phase space

"Mathematics associates new mental images with ... physical abstractions; these images are almost tangible to the trained mind but are far removed from those that are given directly by life and physical experience. For example, a mathematician represents the motion of the planets of the Solar System by a flow line of an incompressible fluid in a 54-dimensional phase space."
Yuri Manin, b. 1937

Into phase space
To understand how attractors (and many of the diagrams used to illustrate chaos) are often represented, we need to take a trip into something known as phase space. Mathematicians and physicists often represent something that is happening in the real world in an imaginary mathematical one where, for example, you might have a dimension for everything that can change in the real world. These dimensions do not represent space as we would usually expect them to, but rather the values that the properties of the object can take.

We can use the familiar pendulum as an example. Its real-world motion is just side to side. But there are a number of ways of representing this movement mathematically. A familiar one, often used in physics, would be to have one dimension represent the position of the pendulum's bob in space and another to represent its position in time. Strictly speaking, such a representation should be four-dimensional, showing all three dimensions of space and one of time—and that's no problem for mathematicians, who can work in as many dimensions as they like. However, to keep it simple, we can often ignore two of the spatial dimensions, and the movement of a basic pendulum can be captured by simply recording the left-to-right position of the bob against the passage of time.

The result is to produce a sine wave—a familiar curve that waves from side to side as it moves forward through time. But this is not the only way to project what is happening into mathematical space. The term "phase space" takes in all possible ways of representing what is happening to a system by assigning different properties to the dimensions of the space. Often, two-dimensional phase space is used to represent the properties of position and momentum (mass times velocity), two key measures in the physics of moving systems. Each point on a phase space diagram then represents the values of position and momentum of a system at one moment in time. So, for example, a phase space diagram of the pendulum could plot the momentum of the bob against its position in space. Each of these properties smoothly shifts from positive to negative as the pendulum changes direction and position. (The mass doesn't change, so the momentum simply follows the velocity, which would be represented as, say, positive traveling left to right and negative traveling right to left.) The result is an elliptical phase space diagram, though it is often made circular by choosing appropriate units.

Of course, real pendulums, without being given a regular push, don't carry on swinging forever. As a result, a more accurate phase space diagram of a real pendulum would be one that very gradually spirals in toward the center of the diagram. That point at the center is an attractor. It doesn't matter how much you disturb the motion of the pendulum by giving it a push partway through the process—it will still, losing energy under gravity, eventually head for that point in phase space.

In practice, many phase space diagrams need to take on more than two dimensions as there are more parameters needed to represent what is happening. Up to three dimensions, we can reasonably plot out a visual representation, but beyond that we've got a problem. One way to add in an extra dimension is to use color—so an extra dimension is added to a point by its position on the spectrum. The result can be impressively beautiful diagrams.

Pendulum in phase space
The pendulum undergoes side-to-side motion. The time series plots position or velocity against time, while the phase space diagram here shows momentum against position.

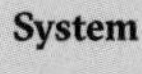

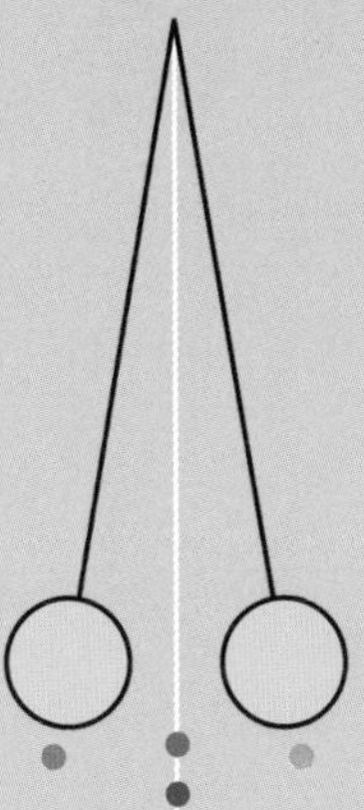

Time series

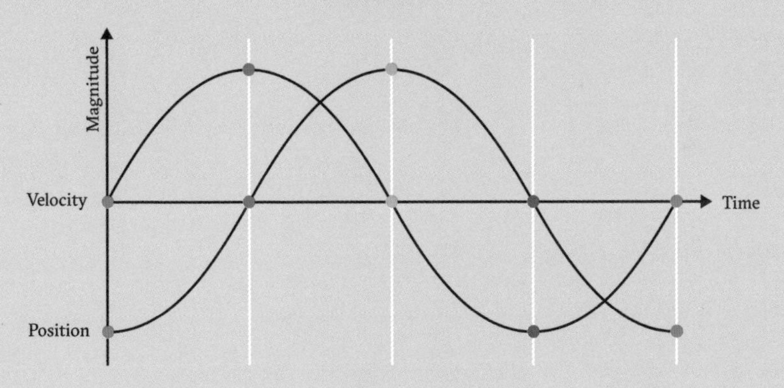

Phase space diagram

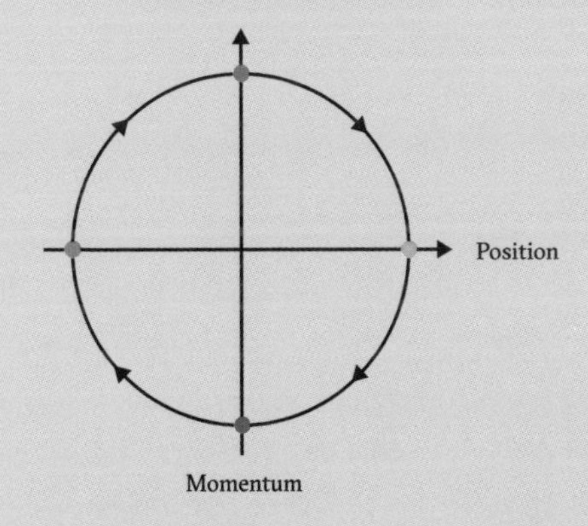

Phase space diagram of real pendulum
A real pendulum will gradually lose momentum, producing a phase space diagram that has an attractor at the center.

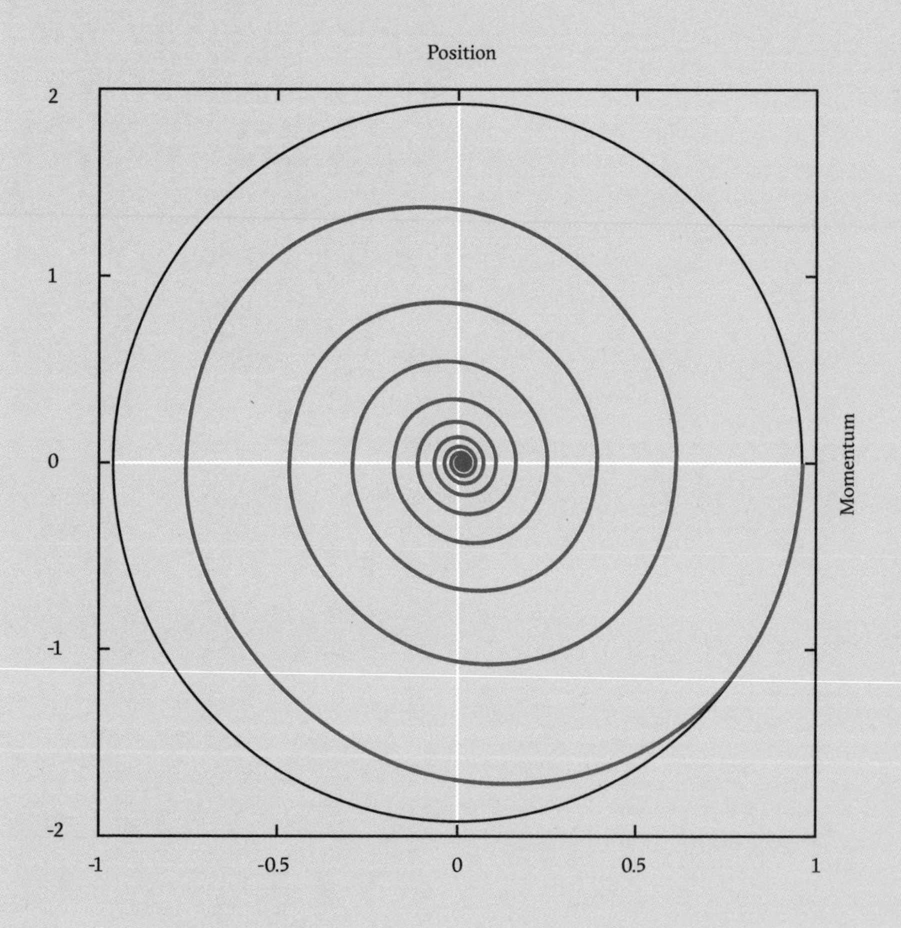

Again, in practice, with most phase space diagrams of chaotic systems, three dimensions aren't so much given a three-dimensional view, but a flat, mapped projection, rather in the way the three-dimensional spherical surface of Earth is projected onto the flat surface of a map. Although we tend to use one familiar projection for maps, there are many other possibilities—and similarly the phase space diagrams used in chaos tend to be a specific slice through the three-dimensional structure that represents one of many possible projections.

More butterflies in the attractor

When a system has one or more strange attractors it does not travel toward them in a systematic way but not appearing to travel in a systematic way toward them, as is the case with the slowing pendulum. Instead, the strange attractor's existence in phase space involves a convoluted path that never crosses the same point twice. If it did revisit a point, it would settle down into a regular motion. Instead there is infinitely fine variation in phase space. Like the Koch snowflake (see below) that led to the idea of fractals, the phase space diagram is infinitely long but occupies a finite area.

Perhaps the best known of the strange attractors is the Lorenz attractor, which emerged from a model that Lorenz was developing for the atmospheric effect of convection (where heated air rises and flows away from the source of heat, carrying heat with it). Where the flow tends to be relatively low in turbulence, one attractor tends to form. As the air becomes more turbulent (with an increase in a factor of fluid flow known as the Rayleigh number) there tend to be two attractors. With the appropriate projection these produce a phase space structure that can be somewhat butterfly-like in appearance—appealing to those who recall Lorenz's butterfly effect.

Another outcome of some dynamic systems can be points of strange repulsion rather than attraction, in some cases producing dramatic fractal forms known as Julia sets, named after French mathematician Gaston Julia.

The best-known name in this field, though, became involved not through considering the weather or working on abstract mathematics, but by studying the behavior of the market price of cotton.

The Lorenz Attractor
Many of the phase space projections of this archetypal chaotic attractor, discovered by Edward Lorenz, appropriately resemble a butterfly.

Self-similarity

"Geographical curves are so involved in their detail that their lengths are often infinite or, rather, undefinable. However, many are statistically 'self-similar,' meaning that each portion can be considered a reduced-scale image of the whole."
Benoît Mandelbrot, 1924–2010

Cotton price distributions
The behavior of financial markets, specifically the price of cotton, was studied in the 1960s by Polish-born, French American mathematician Benoît Mandelbrot. The general assumption from economics at the time—which appears to be common sense—was that you would see a blend of a longer-term impact from major external factors—the economy, technology, financial crises, wars, and fashion trends, for instance—and a shorter-term random distribution of the movements, up and down, of prices, which was expected to follow the kind of bell-shaped normal distribution we encountered earlier. However, in the markets something different was happening.

This had first been noticed by Dutch American economist Hendrik Houthakker. It was picked up on by Mandelbrot when Houthakker invited him to give a lecture. Because cotton was an old market, Mandelbrot was able to go back over more than a hundred years, accumulating lots of data. Achieving clarity can be an issue in the short term with chaotic systems, precisely because of the sudden unexpected movements that typify such a system. Having a small amount of data can give a highly misleading picture. Lacking the economists' mindset, Mandelbrot was able to take a step back from the expectation of externally driven long-term trends overlaid with random noise.

Noise
We are used to the term referring to unwanted or unpleasant sound, but mathematicians, economists, and engineers refer to unpredictable background changes in a series of values as noise.

Mandelbrot found a behavior in the cotton prices that was closer to what is known as a Levy distribution than a normal distribution. A Levy distribution is often found, for example, in the light frequencies given off by heated materials; it has a sudden, sharp off-center peak, representing the fact that a small number of large-scale jumps have a dramatic impact on the outcome. Economists had ignored the random fluctuations as noise in the system, assuming that they did not indicate anything significant about the system itself. Mandelbrot, however, realized that the "noise," particularly these sudden dramatic spikes, reflected the fundamental behavior of the system.

What is more, because of the large amount of data he had available, Mandelbrot was able to observe a particularly strange aspect of the price *changes*, as opposed to the actual prices—something that would prove typical of chaos. When he looked at the overall shape of a plot of the changes, whether he was reviewing a relatively short space of time or many years, the visual appearance was nearly identical. The plot of price changes was, in the jargon, "self-similar." You could superimpose the plot for, say, monthly changes on that of the daily changes and pretty much they matched. All the external factors that got the economists so excited made hardly any difference.

At the heart of the messiness of the way prices jumped up and down was a hidden structure. This reflects the chaotic nature of the factors controlling financial markets. As we will often discover, this doesn't mean that Mandelbrot had unwrapped a magic bullet to predict the behavior of the stock market. The uncovered pattern would not enable anyone to buy or sell early and make a fortune. But it did show a totally different working in this kind of system to that assumed by economists.

Predictably unpredictable

Mandelbrot was not an economist—in fact, it would be hard to pin down exactly what he *was* to any one discipline. However, he spent most of his working life at IBM's research headquarters, the Thomas J. Watson Research Center,

Levy and normal distributions

The more familiar normal distribution (bottom) is symmetrical with a central peak, whereas the Levy distribution has a large off-center peak where a small number of large-scale events influence the outcome. Both c for the Levy distribution and α for the normal distribution describe how widely spread the values plotted are.

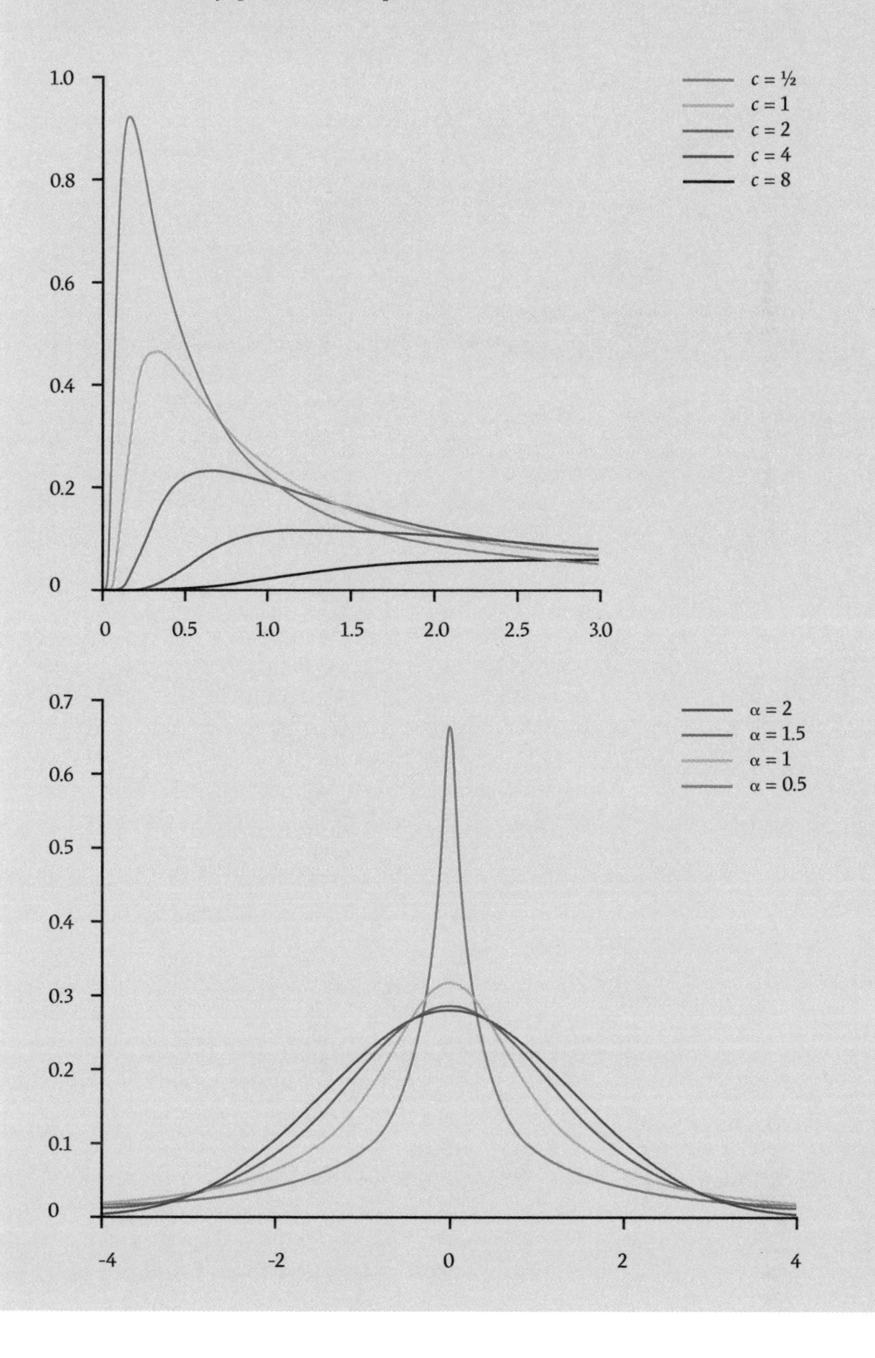

located in Yorktown Heights, New York. Like a number of large organizations including Xerox and 3M, IBM was prepared to allow its researchers freedom to spend at least some of their time exploring whatever took their fancy, knowing that this unrestrained research was often the way that the most far-reaching breakthroughs occurred.

As a result, Mandelbrot's work ranged free and only occasionally overlapped with IBM's core businesses of computing and communications. Once Mandelbrot had spotted the strange self-similar behavior in cotton prices, he started seeing it regularly in systems—and one instance would give a positive benefit to his employer.

In the 1960s it was starting to become common to use telephone lines not only to carry voice communication, but data too. When you're talking to someone on the phone it doesn't really matter if there are some random fluctuations in the signal, producing noise in the background, but it becomes more than an irritant when trying to transmit data. IBM engineers were looking for ways to minimize the impact of such noise on their communications lines.

The problem was that the engineers could not find any pattern in the data that would enable them to get a handle on it and try to eliminate it. Mandelbrot, though, approached the data from another angle. He split it up into chunks of time, some of which contained noise and some of which didn't. Looking at a chunk with noise, he found it too could be split up into chunks with noise and chunks without. And the same split went on as he continued to zoom in on the data. It was self-similar in the sense that whatever level Mandelbrot looked at it there were always chunks with noise and chunks without—it was never possible to pinpoint purely noisy parts of the data.

As a result, Mandelbrot was able to advise the engineers to take a different strategy. When a burst of noise arose, their usual reaction would be to look for a cause—something that had been changed in the system or its environment to distort the signal—yet it was clear that this was a chaotic structure that didn't require a cause for the noise to arise. And it was also clear that simply making the signal stronger would not overwhelm the noise, but rather would generate more noise. As a result, Mandelbrot could direct the engineers to look for alternative ways to live with the noise.

This discovery was to underscore IBM's approach to data visualization. However, it was not the work that was to make Mandelbrot's name. This started instead with a very familiar chaotic environment: the convoluted twists and turns of a coastline.

The coastline conundrum

In 1967, Mandelbrot published a paper that began an exploration of self-similarity, the driving principle of what would become known as fractals. His particular example was the coastline of Britain. Any atlas or encyclopedia will specify the distance around a coast—but what does this actually mean? Mandelbrot imagined traveling along a path around the entire coast, measuring progress with a yardstick and coming up with a definitive distance.

Self-similarity
A fractal shape like a coastline is self-similar in that a small section of it, blown up, is visually similar in its twists and turns to the original.

Let's say that the distance measured was 3,000 miles (4,800 kilometers). Now set off along the coast path again, this time measuring using an inch-long measure because it will go into more fine detail of the twists and turns of the path than the yardstick. This time, the distance might be 4,000 miles (6,400 kilometers). Then, try it again, laying a piece of string around the cliffs and beaches, which would be able to find its way into far more small chinks and crevices. The distance now would be even longer. Depending on the measurement approach used, Britain's coastline has been given as anything ranging from 1,700 miles (2,800 kilometers) to 11,500 miles (18,500 kilometers).

It's possible to imagine taking this approach to greater and greater extremes, measuring every single indentation caused by the individual atoms on the surface of the rocks around the coast. Which measure is the "real" distance? There isn't one: the only answer can be any and all of them. That is the paradox that Mandelbrot made mathematically explicit. There is not a distinct, definable distance around a coastline (or along any other crinkly structure, for example a border between neighboring countries or the length of a river). However, it is possible to provide a mechanism to explain what is happening, and to do this, Mandelbrot dived into the mind-bending concept of fractional dimensions.

Measuring the coastline of Britain

This shows how decreasing the length of a measuring stick gives a more accurate, longer value for the length of the coastline. In the examples, N is the number of measures in the perimeter and r is the fractional size of the measure (so $r=2$ is half the size of $r=1$).

Enter the fractal

"I conceived and developed a new geometry of nature and implemented its use in a number of diverse fields. It describes many of the irregular and fragmented patterns around us, and leads to full-fledged theories, by identifying a family of shapes I call fractals."
Benoît Mandelbrot, 1924–2010

Counting dimensions
We are used to dimensions coming in whole number form. The familiar space of the real world has three dimensions. An image on a sheet of paper has two dimensions, and an isolated straight line is one-dimensional. We also know that mathematicians happily work with many more dimensions in imaginary spaces. But what would it meant to have, say 1½ dimensions? The concept of a different kind of dimensionality where this could arise had occurred to English mathematician Lewis Fry Richardson in the early part of the twentieth century. A pioneer of the use of mathematics in weather forecasting, Richardson had been working on an entirely different topic: trying to understand the effect of borders on influencing countries to declare war.

Richardson noted that the border between Spain and Portugal was declared by Spain to be 613 miles (987 kilometers) long, whereas Portugal asserted it was 754 miles (1,214 kilometers).

While wondering if this disagreement had sparked conflict between the countries in the past, Richardson discovered a relationship between the length of the measuring stick used and the distance produced that was controlled by a factor he referred to as the border's dimension—a figure that in this case had a value between 1 and 2. It seemed that there was an intermediate nature of the dimensionality: something that lay between a one-dimensional straight line and a two-dimensional shape.

At first sight, the concept of a non-integer dimension might seem bizarre, but a useful way of looking at fractional dimensions is to think of traditional dimensions not as directions in space, but as the smallest number of ways something can be divided up to make sets of identical miniature versions of itself. A line can be split into two equal-length lines. A square can be split into four identical squares. A cube divides into eight smaller cubes. (Of course, in each case you can split the original shape up smaller, but these are the least possible divisions.) The number of divisions is $2n$, where n is the number of dimensions. When we move on to shapes with fractional dimensions, the fractional dimension effectively reflects the minimum number of smaller copies which can be fit together to make the full object.

The concept that preoccupied Richardson was developed by Mandelbrot to define a specific ratio to show how the detail in a pattern changes (or doesn't) over the scale by which it is measured—the size, if you wish, of the measuring stick. For simple shapes, what Mandelbrot initially called a fractional dimension, later a fractal dimension, is exactly the same as the familiar dimensions, but for shapes that have self-similarity, like a coastline and borders, it goes beyond whole number values. As the fractional value comes closer to an integer it resembles more what we would expect a shape with the integer dimension to be like. So, for instance, a curve with a value just above 1 is pretty much like a straight line, while one that comes near to 2 has significantly more complexity.

One important point to note here. This isn't tinkering at the margins. The math we are taught at school and the simplifications of physics both often impose simple, straight-line shapes on the world, but the real world is usually crinkly. These intermediate dimensions are the norm in nature, not an oddity. This work inspired Mandelbrot to come up with a new term which we have already encountered several times, but now can take the plunge and fully explore: It was Mandelbrot who coined the term "fractal" in 1975.

Snowflakes, gaskets, and sponges

Mandelbrot's conception of a fractal was anything where the fractional/fractal dimension was greater than its more familiar "topological dimension"—one for a straight line, two for a surface, and so on. His term was apparently derived from *fractus*, the Latin for broken (the same source as fractured). In appearance, a fractal has the kind of self-similarity Mandelbrot had noted in the coastline of Britain and in the noise on an IBM data transmission line. Zoom in and look at a fractal in greater and greater detail, and the result is that it continues to exhibit the same amount of ruggedness; it isn't smoothed out by zooming in or out in view.

Topological dimension
Topology is the field of mathematics concerned with the properties of shapes that can be stretched and distorted to any degree, as long as they are not broken or merged.

Although Mandelbrot coined the term and opened up the concept, he wasn't the first to explore some of the oddities that can arise from what we now know as fractals. One of the earliest examples to be studied was the Koch snowflake, named after Swedish mathematician Helge von Koch who devised it in 1904.

The Koch snowflake starts with a simple equilateral triangle and is produced by dividing each straight line in the shape into thirds and constructing a smaller equilateral triangle on the middle third of each line, pointing outward. This process is then repeated over and over. The shape goes from being a triangle, to a six-pointed star, to a more and more crinkly snowflake shape that exhibits the classic fractal nature of being self-similar: If we zoom in to a part of it to discover greater detail, the section perfectly reflects the larger-scale image. The Koch snowflake has a fractal dimension of around 1.26.

Perhaps most remarkable is the distance around the snowflake. Although the snowflake never expands outside a circle drawn around its original three points, the perimeter of the snowflake gets longer and longer. Despite the lengths of the newly crinkled sides getting shorter and shorter, the total length tends to infinity as the number of iterations increases. This is because the total perimeter size is three times the original side length multiplied by four divided by three raised to the power of the number of iterations. This is a number that tends to infinity as the number of sides increases. We get an infinite perimeter, despite the snowflake having a finite area.

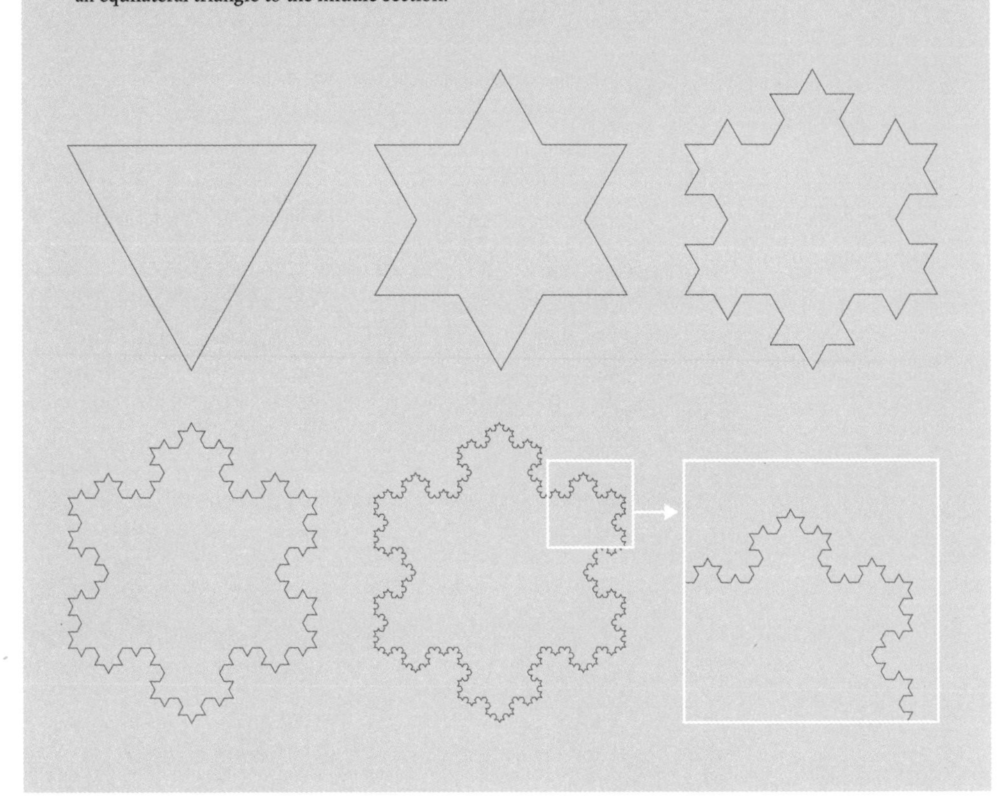

Koch snowflake
The first five iterations of the Koch snowflake all involving splitting each line into three and adding an equilateral triangle to the middle section.

An equally impressive, yet still simple, fractal shape is the Sierpiński gasket, named after Polish mathematician Wacław Sierpiński, who devised it in 1915. Rather than adding sides to a line, the gasket is constructed by removing areas from a shape. It starts again with an equilateral triangle, but here we consider the triangle to be a filled-in shape rather than an outline. We start by removing an inverted equilateral triangle with points in the middle of each side of the original triangle. The process is then continued by removing triangles in the same way from each of the filled-in triangles left by the previous step.

Although the resultant shape is sometimes called the Sierpiński triangle, the term "gasket" is a better description as the important aspect of the fractal is its construction as a series of holes cut out of a sheet. Apart from appearing attractive, the gasket is self-similar, with a fractal dimension of around 1.59.

Sierpiński gasket
After each iteration, the gasket has less surface area, tending to zero. Any upward-pointing triangular section is self-similar to the whole.

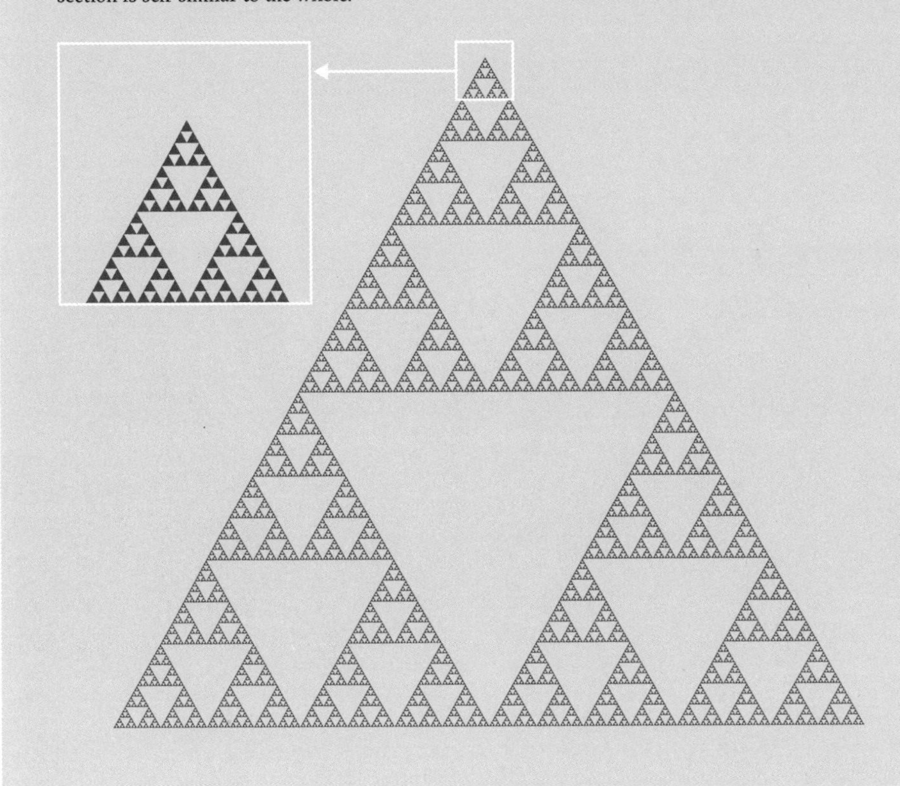

Like the Koch snowflake, the sting in its tail lies in the limit as the process is repeated more and more: although the "material" in the gasket never all disappears, at the limit it has an area of zero.

Austrian American mathematician Karl Menger enhanced a two-dimensional variant, known as the Sierpiński carpet, created by dividing a square into nine and cutting out the middle square. The process is then repeated for each smaller square—and so on. Menger produced a three-dimensional equivalent in what is now known as the Menger sponge which, taken to the extreme, contains no volume of material yet has an infinite surface area.

These and other relatively simple fractals demonstrate the typical chaotic evolution of complexity from small variations—but neither the snowflake nor the gasket is as archetypal of a fractal as the design known as the Mandelbrot set.

The Mandelbrot set

Exactly who first came up with the Mandelbrot set is a matter of dispute. Mandelbrot always claimed it for his own, but an early version of it appears in the work of two American mathematicians, Robert Brooks and Peter Matelski, from 1978, while it seems a French/American pairing in the form of Adrien Douady and John Hubbard could be responsible for naming it in Mandelbrot's honor.

Whoever devised it, what we have here is the literal poster-child of fractals and chaos, appearing on everything from framed artworks to tablemats. In the beautiful and complex shape, which seems to mix images of bizarre living organisms with paisley curves, is a structure that features self-similarity, but in a much more intricate and exotic fashion than snowflakes and gaskets. The set has been described as the most complex object in mathematics, a mind-boggling mix of curves and spikes, spirals and medallions, which in its colored visual representation became a new form of art. This move from pure mathematics

←

Menger sponge
After four iterations, this three-dimensional equivalent of a Sierpiński gasket is already highly self-similar.

to visual wonder reflects the timeliness of Mandelbrot's position at IBM, then a center of computer graphics.

In ancient Greek times, mathematics had been all about drawings. Numerically, their math was very limited (not helped by the way that they used letters as numbers) and did not have any way of properly representing fractions or formulae. As a result, ancient Greek mathematics was dominated by geometry. Over the years since, mathematics had become ever more symbolic, moving away from pictorial representation. But with the new peculiarities Mandelbrot was uncovering, the ability to access IBM's technology at the leading edge of the computer graphics of the day meant that he could bring a new and fascinating visual approach to chaos.

The exact formulation of the Mandelbrot set is a little messy, though the rules for constructing it are simple. At its heart is a set of complex numbers. These are numbers combining an ordinary "real" number, such as 1.3524, with an "imaginary" number—which is a real number multiplied by the square root of -1. The square root of -1 has no value in the real world, because 1 × 1 = 1 and -1 × -1 = 1—there is nothing that multiplied by itself is equal to -1. However, for several hundred years, mathematicians have made use of this concept, represented by *i*. When multiplied by itself, *i* × *i* = -1. Initially an imaginary number was little more than a mathematical oddity, but it was realized that the combination of a real number and an

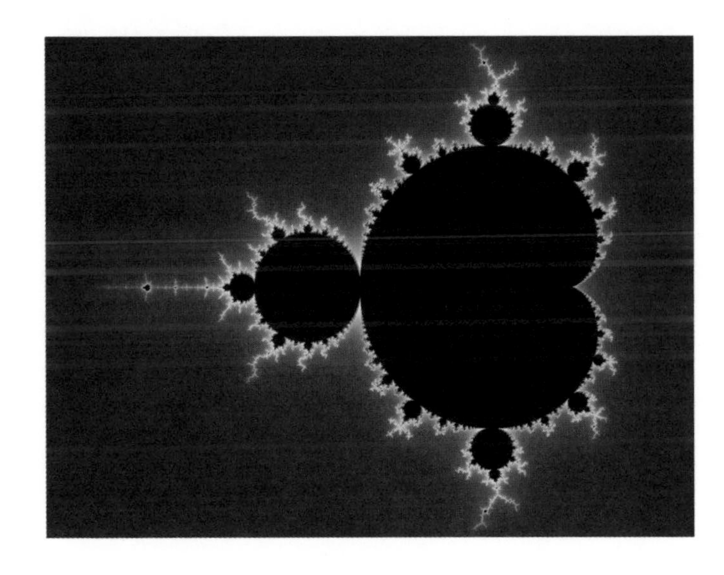

→

Mandelbrot set
The overview plot of the whole Mandelbrot set—zooming in produces a range of dramatic, self-similar shapes.

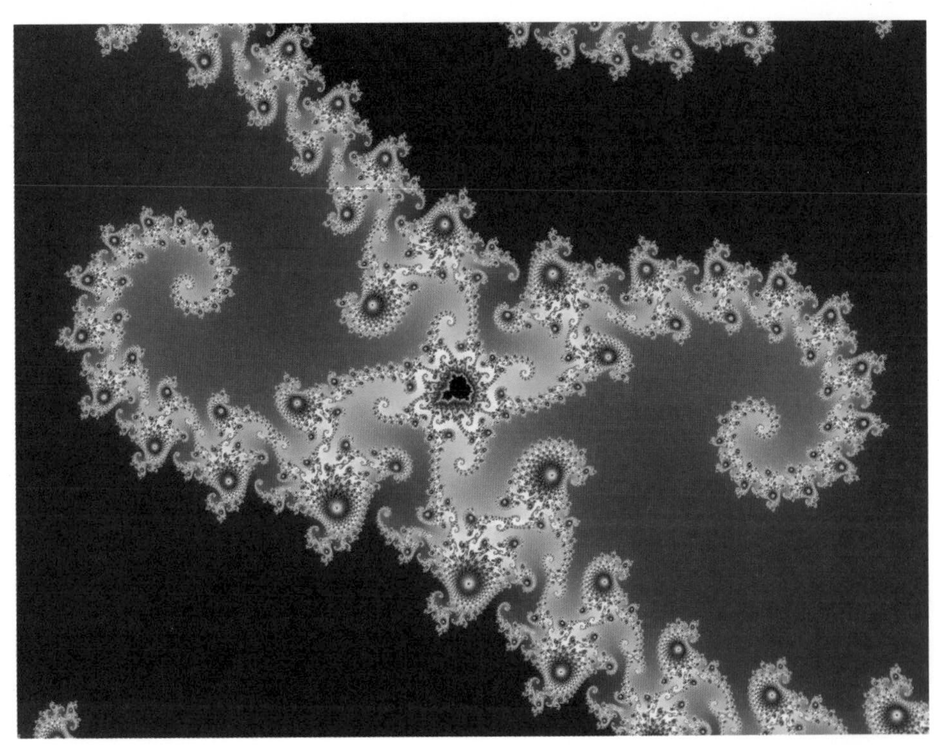

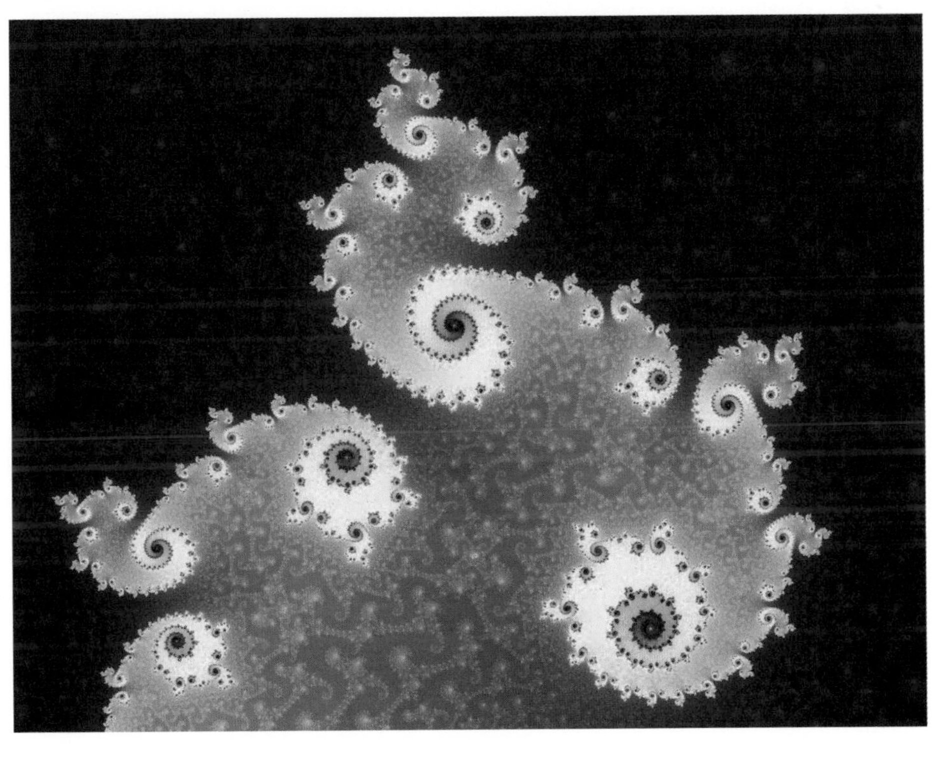

Mandelbrot set detail
Zooming in to a small section of the Mandelbrot set results in a wide variety of features.

imaginary number such as 1.316 + 2.64*i* could be used to represent a point on a two-dimensional plane. Known as complex numbers, these values are ideal for representing shapes that evolve over time, such as waves, and so became invaluable to physicists and engineers alike.

The Mandelbrot set is a set of complex numbers, generated iteratively, like the snowflake and the gasket, but through a slightly more arduous mathematical process. The result is to identify a series of points that are in the set, which are then plotted out to produce the classic Mandelbrot set images. One way to calculate the set is to take any point in the complex plane, then repeatedly square that value and add it to the original value. If the outcome heads off to infinity, the point is not in the set—if it is finite, it is in the set. So, for example, 0.4 + 0.1*i* disappears to infinity, so is not part of the Mandelbrot set, but 0.3 + 0.1*i* oscillates around, eventually heading toward a small finite value, and is in the set. Points that are in the set produce a loop of values, or a chaotic series of jumps up and down that never head off toward infinity.

The basic set generates black and white images, but the truly beautiful Mandelbrot set images that would come to hang on so many walls are multicolored. In mathematical terms, the colors don't really signify anything hugely valuable—they just make the patterns more attractive. Remember that the deciding factor of whether or not a point is in the set is the outcome of repeatedly squaring a value and then adding the initial value. Multicolored Mandelbrot sets simply reflect the number of iterations required to pass some target value. Delightfully, the self-similar nature of the Mandelbrot set reproduces by the same factor as the Feigenbaum number (see page 119) discovered for period-doubling branching. Each of the "blob" shapes found in different levels of the set, for example, is about 1/4.669 times the size of the previous one.

Feigenbaum number
As detailed earlier, the Feigenbaum number is a constant of nature, approximately 0.1242, reflecting the reducing size of the factor causing the split in period doubling.

Although there are some similarities to natural shapes in the Mandelbrot set, other explorations of fractals showed that coastlines were not an oddity. Many shapes in nature have distinct fractal qualities.

The fractal universe

"Fractal geometry will make you see everything differently. There is a danger in reading further. You risk the loss of your childhood vision of clouds, forests, flowers, galaxies, leaves, feathers, rocks, mountains, torrents of water, carpet, bricks, and much else besides. Never again will your interpretation of these things be quite the same."
Michael Barnsley, b. 1946

Fractal nature
Because of the way that they form, a range of natural phenomena tends to be approximately fractal in appearance. Perhaps best known are some kinds of plants, particularly evergreen trees and ferns, where there is self-similarity between a branch and the plant as a whole. In a more sophisticated fashion, mountain ranges and clouds both tend to have fractal forms where, like a coastline, zooming in to their edges produces more and more crinkles of detail. Even an apparently smooth real-world object will tend to have fractal edges as you zoom in at high magnification.

Because of this fractal appearance, visual effects software, such as the code used in games, has often made use of fractals in the generation of landscapes and cloudscapes. The result is an impressively realistic-looking scenery which can be generated from a small part of the program.

Romanesco
The brassica Romanesco exhibits impressive self-similar fractal structure.

Perhaps most visually impressive of all the natural fractal forms occurs in a relation of cauliflower and broccoli known as Romanesco. This remarkable brassica is so dramatic in its self-similar structure it looks more like something produced by an imaging computer than any natural growth. Although the Romanesco was not specifically cited, it's likely that examples like this were among the inspirations of one of the few practical applications of fractals that go beyond producing computerized landscapes and arty designs: compressing images.

Fractal compression

It can be hard to remember now, but back in the 1990s, the personal computer industry had a significant storage crisis. The amount of data being held (particularly once computer-based photographs started to be used) was expanding far faster than disk drive capacity. A main reason for this was the rapidly increasing resolution of computer graphics, meaning that those images took up more and more of the available space. As digitization of images and digital photography was taking off, this presented the industry with a serious problem.

It was quite common at the time for computers to run disk compression software that made it possible to pack more data into the same-sized disk drive, and although this cleverly worked behind the scenes without the user having consciously to do anything, the software slowed down programs due to the time taken to compress and decompress the data. What was ideally required was a way to compress images as much as possible, storing them in far less space, that was nevertheless extremely quick to decompress. The time it took to compress the image initially was less of an issue, because this was only done once. It was here that fractal compression briefly shone.

Devised originally in 1987 by English mathematician Michael Barnsley, by 1992, fractal compression had become available. This searched for self-similarity in images to make use of the fractal nature of real-world objects, compressing images to take up less room. The software was capable of compressing pictures—especially those of natural features—much smaller than conventional compression techniques for the same level of detail, at the cost of taking a long time to process the initial compression.

Briefly, it looked as if fractal compression would become the new norm. It was used in some computer games and in

Microsoft's groundbreaking *Encarta* CD-ROM encyclopedia, which was significantly responsible for the demise of the paper encyclopedia before Wikipedia showed up. But before long, new disk technology rendered fractal compression redundant. JPEG compression proved more than enough for the job—and quicker to use. A JPEG image might be ten times the size of a fractally compressed image for the same quality—but with disk space to spare, this didn't matter. Fractal compression came and went—but chaos in the broader sense was here to stay. And nothing makes this clearer than when we look at some of the ways that traditional forecasting has failed far more than was the case with the weather.

5
Stock Market Crashes and Super Hits

Misusing probability

"There is a special department of Hell for students of probability. In this department there are many typewriters and many monkeys. Every time that a monkey walks on a typewriter, it types by chance one of Shakespeare's sonnets."
Bertrand Russell, 1872–1970

Monkeying with Shakespeare
The great English philosopher Bertrand Russell summed up pithily the old idea that a roomful of monkeys, randomly hitting the keys of typewriters, would reproduce the works of Shakespeare—given enough time. This quote is partly used to illustrate our faulty grasp of what randomness looks like. Ask someone to type a long random string of letters and they will tend to produce far fewer repeats of letters, or near-words, within the string than true randomness requires there to be. Common sense encourages us to misuse and misinterpret probability.

To be fair, that traditional roomful of monkeys, which predates computers and random number generators, is not a great way to provide a random sequence of letters, but as we have seen, it is perfectly possible to generate such a sequence. Let's take just the opening line of one of Shakespeare's best-known sonnets, Sonnet 18: "Shall I compare thee to a Summer's day?" If we make it all uppercase, and ignore all punctuation except spaces, we need 27 characters. This means that the chance of a genuine random "roomful of monkeys simulator" coming up with an S is 1 in 27. That's not too bad. The chance of generating SH is 1 in 27×27, or 1 in 729. This is admittedly already relatively unlikely for one attempt at typing a pair of letters, though we should bear in mind that SH is just as likely to turn up as any other two-letter combination.

In principle, then, the suggestion that the monkeys would eventually type Shakespeare seems feasible. However, the chances of getting that opening line alone is 1 in 27^{37}. That's about 1 in 10^{51}—1 followed by 51 zeros. Compare that with the chance of winning one of the major U.S. lotteries, which is typically about 1 in 3×10^8. Getting just that line typed is a pretty unlikely outcome. Of course, people do win the lottery and in principle the monkeys could also produce that sentence—but it is ridiculously improbable. With 1,000 monkeys, typing 37 characters in a generous 10 seconds, it would take on average 10^{47} seconds, which is about 3,000 trillion trillion trillion years. Russell was emphasizing in his comment about hell just how much the monkeys and typewriters illustration misuses everyday probability. Chaos presents the same kind of risk to the unwary, where probability seems to offer a reasonable way to understand something, but when you drill down to the detail, it is likely to lead you astray.

Bookies' odds versus the casino

As we have already discovered, the origins of probability were tightly linked to gambling games, where the likelihood of a particular outcome is controlled by probability (assuming a fair game). So, for example, we can reasonably expect a 1 in 2 chance of winning on the toss of a coin, a 1 in 6 chance on the roll of a die, and a 1 in 13 chance of a particular card being drawn from a pack, ignoring suits. Similarly, on the spin of a roulette wheel, if we ignore the 0 where the house always wins, there's a 1 in 2 chance of guessing red or black correctly or a 1 in 36 of being correct with a specific number.

POLLING PLACE
VOTE

←
Election odds
Unlike the predictable randomness of a coin toss, odds on the winner of an election are based on attempting to predict the outcome of a chaotic system.

Outside of casino-type games or lotteries, where there is a specific probability of any particular outcome because a random process is used to generate it, most betting is dependent on the odds offered by a betting shop or bookie. It's easy to be fooled into thinking that this is the same kind of exercise, because again we are offered odds on a particular outcome—whether it is the result of a horse race or an election—but in reality what appears to be driven by a similar kind of randomness is, in fact, reflecting an attempt to second-guess the outcome of what is usually a chaotic system. This is why, for example, bookies were offering surprisingly good odds for those willing to bet on Donald Trump winning the presidency or the U.K. leaving the European Union. The odds did not represent the true probabilities of these events occurring, rather, a failed attempt to impose probability on systems that had chaos at their heart.

Poll dancing

Polls and studies are a common feature of modern life. We are always being told the outcome of some questionnaire or other, with the numbers inflated to try to reflect a whole population. So, for example, rather than tell us that out of 1,000 U.S. citizens polled on the internet, 50 percent supported some cause, we might be told that 164 million Americans (half the population at the time of writing) take this view. This is the result of using probability and statistics, attempting to draw inferences about a wider population based on a sample.

The biggest problem with this approach is ensuring that the sample is truly representative—and even then there can be issues, either because the sample is representative but the measure isn't (we'll unpack that in a moment) or because the system is sufficiently chaotic that the probabilistic approach of sampling simply doesn't work well enough to predict an outcome.

Let's start with a representative sample where the measure chosen isn't representative. A good example of this is the Public Lending Right (PLR) system in the U.K. PLR is a mechanism used by a number of countries to pay authors a small fee each time one of their books is borrowed from a library—much like the way that Spotify pays musicians a very small amount each time one of their tracks is streamed. Spotify is a modern, big data system, so collates information on every single stream. However, the pre-big data PLR technique samples a number of libraries intended to be representative—they are selected for

Demographics
The study of a human population and its characteristics; the makeup of a particular population in anything from ethnicity to education.

their mix of stock and the demographics of their readership—and scales the results up for the system (in this case, the country) as a whole.

This would be fine if all books were equally attractive in all libraries. But let's say I write a book about my hometown. This will be borrowed far more from libraries in that town than it will be elsewhere. If one of my town's libraries is not in the (relatively small) set of libraries sampled, borrowings of my book will not be reflected. (Equally, if the book *is* in the sample, borrowings will be greatly exaggerated.) The libraries chosen for the sample are representative of U.K. libraries when considering all borrowings, but that measure is not representative of the specific subclass of books on local topics.

As for sampling from a chaotic system, if the topic is one where opinions are relatively strong, then there is not too much of a problem. If you ask me whether I prefer to eat meat or tofu, I have a very clear answer, and it is not something that is likely to be swayed by argument. However, if you ask me which political party I am likely to vote for, the outcome is much more swayed by a system with a range of interacting factors: the parties' policies at the moment, who my local representative might be, what I think of the different parties' leaders, and more: It is a classic chaotic environment. This isn't true of everyone—some vote reflexively, always voting the same way, however good or bad the different options are. But for the set of floating voters, which in many countries is increasing in size, the nature of this chaotic system is crucial.

Whatever the type of system being studied, there will be problems unless the sample is representative of the population as a whole—and this is extremely difficult to untangle. The only kind of poll that isn't self-selecting is a compulsory one. With a voluntary poll—which is pretty much all of them except censuses and voting in some countries—often the mechanism of taking the poll itself biases the sample. So, for example, the majority of modern polls are taken online. That immediately provides a bias on age, technical experience, and more. Similarly, polls taken over the telephone or by stopping people in the street will tend to select a particular demographic. Even the time of day such a poll is taken will have an influence. And that's just the potential for bias in the means of gaining the information. The size of the sample will also have an impact—far too many social science studies, for example, are based on numbers of participants that are too small to give any confidence to the findings. Equally, the way the questions are phrased can bias the outcome—even the other questions surrounding a question can do this. And so it goes on.

Poll companies and scientists undertaking studies attempt to remove the bias—but this involves changing the actual findings in a way that is supposed to make the data more representative. This can be both deceptive—the actual poll results may have been totally different from the ones reported—and open to conscious or unconscious manipulation, as the process is inevitably subjective. Polls and small-scale studies may often be our only option—and sometimes they are better than nothing—but we ought to be clearer just how inaccurate they are.

Let's take a look in some more detail at a few specific areas where chaos can mislead, starting with the financial markets, stocks, and shares.

Playing the markets

"I tell people investing should be dull. It shouldn't be exciting. Investing should be more like watching paint dry or watching grass grow. If you want excitement, take $800 and go to Las Vegas."
Paul Samuelson, 1915–2009

The dangerous "because"
The "mainstream media" has come in for significant criticism from politicians lately. In reality, much of this criticism is misplaced, but even the most unbiased of the news media is in danger of getting things wrong when faced with science and mathematics, not helped by the fact that relatively few journalists have a math or science background. Something that repeatedly trips up journalists is confusion over correlation and causality.

Correlation means that two separate things change at the same time or in the same way, whereas causality means that one thing causes the other. We have a natural tendency to assume that correlation implies causality. But it could equally be pure coincidence, it could be that the causality runs in the opposite direction to our assumption (so B causes A, rather than A causing B), or it could be that a third factor has caused both observed things, rather than one influencing the other.

The importance of not assuming that correlation is the same thing as causality can be seen in some of the bizarre correlations that have been discovered by the Harvard Law student Tyler Vigen on his Spurious Correlations website.

Correlation versus causality
Statistically, the number of people who have died from being tangled in their bedsheets is strongly correlated to cheese consumption—but there is no causal link. (Source: Tyler Vigen)

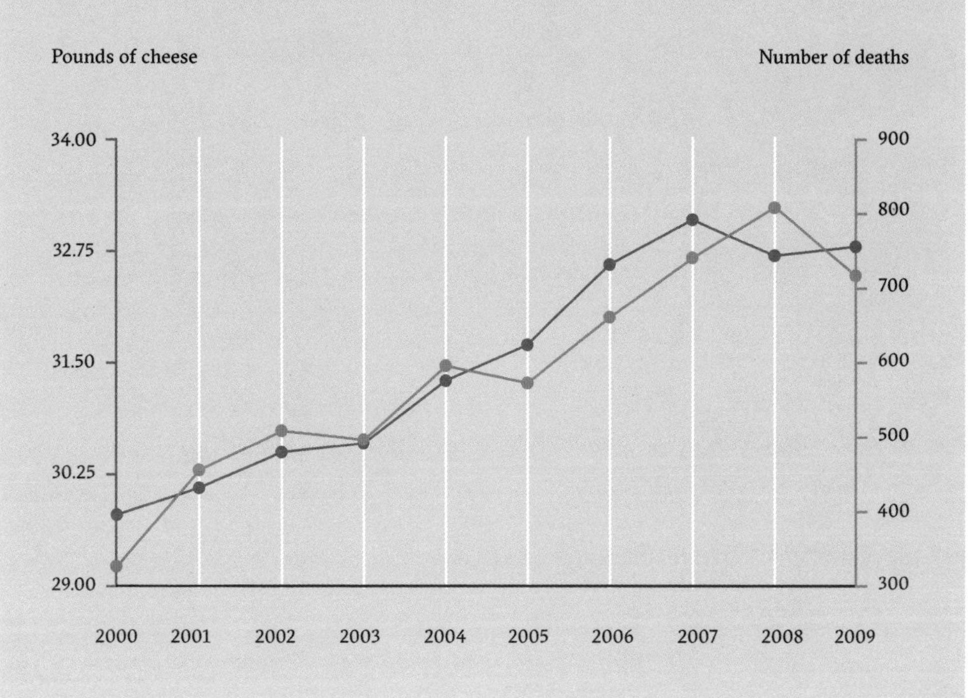

Vigen shows, for example, that there is a very strong correlation between drivers killed in collision with trains on U.S. railroads and the U.S. imports of crude oil from Norway. There is no suggestion of causality here, but in less unlikely combinations, we find it very easy to assume that there is a causal relationship just because things happen together.

Such assumptions crop up regularly in the news, often in terms of the behavior of share prices, where we are told that, say, stocks in a technology company fell "because of a copyright lawsuit" or "because of tough trading conditions" or "because of poor reviews of a new product." In reality, all we know is that there is a correlation between the two. These events occurred at a similar time to the change in the share price. No one can be sure of causality here—there is no simple link between news about the company and how it fares on the market. All we can say is that an announcement was made and at a similar time share prices fell. The word "because," almost always used in this circumstance, is hardly ever justified.

Model behavior

The Nobel Prize winning economist Paul Samuelson, who is said to have made the remark that investing should be dull, wrote a paper back in 1965 entitled "Proof that properly anticipated prices fluctuate randomly." His aim seems to have been to dissuade readers from trying to second-guess share price movements, which is simply a form of gambling, and instead to look for long-term growth across a basket of shares. But there is a little sting in the tail of those words "fluctuate randomly."

Traditional, so-called neoclassical economics teaches those who study it that economies have a kind of inherent self-stability. Of course, there is random noise, but the idea is that any large fluctuations are damped down by the "invisible hand" of the market. The invisible hand was introduced by eighteenth-century Scottish economist Adam Smith to describe the way that an unintended consequence of self-interest is to provide wider beneficial outcomes. As Mandelbrot demonstrated, the behavior of market prices is not true randomness, however, but rather the unpredictability of a chaotic system, and in such a system, it is the sudden jumps that typify behavior, not the underlying trends.

→
Stock traders
Despite many attempts to apply probabilistic forecasting, the movement of prices on a stock market is chaotic.

Few systems better represent chaos than the modern stock market. Although buying and selling shares is a simple enough process, the system involves a wide range of interacting factors. Individual traders, external influences, all the different factors that the news media likes to blame for stock market movements are present and interacting. To make matters worse, because of the nature of modern electronic trading, the timescales for the system to undergo dramatic changes can be fractions of a second.

So, despite many models of the stock market using old-fashioned probability, the reality can be very dependent on initial conditions and can result in extremely rapid deviation from expectation, often involving feedback loops that accelerate the outcome. A vivid example of this was the so-called "flash crash" of 2010, when more than a trillion dollars was wiped off the value of U.S. stocks in just over a half hour. Software trading packages that were designed to make sell decisions based on what happened in the previous minute got themselves into a dangerous positive feedback loop. A probabilistic model forecasting the way that the stock market would move would be totally thrown by the sudden, extreme nature of a crash like this.

Feedback and gaskets

"The four most expensive words in the English language are 'this time it's different'."
Sir John Templeton, 1912–2008

Why be so negative?
If markets are to keep the economy roughly in equilibrium (as assumed by neoclassical economics), it follows that the interactions within the economy must be dominated by negative feedback. In practice, though, there are several positive feedback loops in the economic system. These become clear when we get a boom or a bust—for example in the dot-com boom between around 1995 and 2000. In a boom, particular stocks build a momentum. Because traders see these stocks being bought into, they leap onto the bandwagon.

There are also human parallels to the "flash crash" electronic positive feedback loop, driven by far greater connectivity than used to be the case in market trading. Thanks to social media and other person-to-person electronic communication, it is far easier now for network effects to enter, where interaction between investors results in a movement in one direction being amplified by positive feedback, resulting in the financial equivalent of the screech of a speaker.

An extreme example of dangerous positive feedback is when there is a run on a bank. The bank's equilibrium is dependent on a degree of trust being maintained with its customers. If trust collapses, as the message gets around that the bank is in trouble, people start to withdraw their holdings. This results in further falls in the bank's share price, which encourages more people to get their money out, feeling that there will come a time when the bank will be unable to come up with the cash. Both the bank's liquidity and its share price go into parallel spirals of decline.

So why, then, do economists still pretend that markets are natural, self-correcting, stable environments? It has been suggested that it has a lot to do with economics professionals keeping their well-paid jobs. If they were to admit that economies were chaotic systems where there was no possibility of ever being able to make useful predictions beyond the very short term, there would be a loss of trust and less inclination to take them seriously. Economists have always claimed that economics is a science, but if it is, it's as if physics were still trying to apply a two-centuries-old understanding to the universe when we now know about quantum theory and relativity.

Sierpiński stocks

Canadian mathematician David Orr points out a fascinating resemblance between stock market crashes and the Sierpiński gasket that we met on page 136. Although we can't use traditional probability and the normal distribution to predict what stock prices will do, the kind of chaotic movement we see on the stock market reflects what is known as a power law. This means that for each scale of major change, such as a crash, there will typically be about one-third the number of events as there are for the next lowest scale.

Power law
A power law, such as the law of gravity, is one where one value depends on a power (for example the square or the cube) of another value.

In this respect, the sudden drops are distributed rather as the empty spaces are in a Sierpiński gasket. For each massive drop, there are around three times as many drops of half the size, around nine times as many drops of one-quarter the size, and so on. When the size of the change halves, there are around three times as many events.

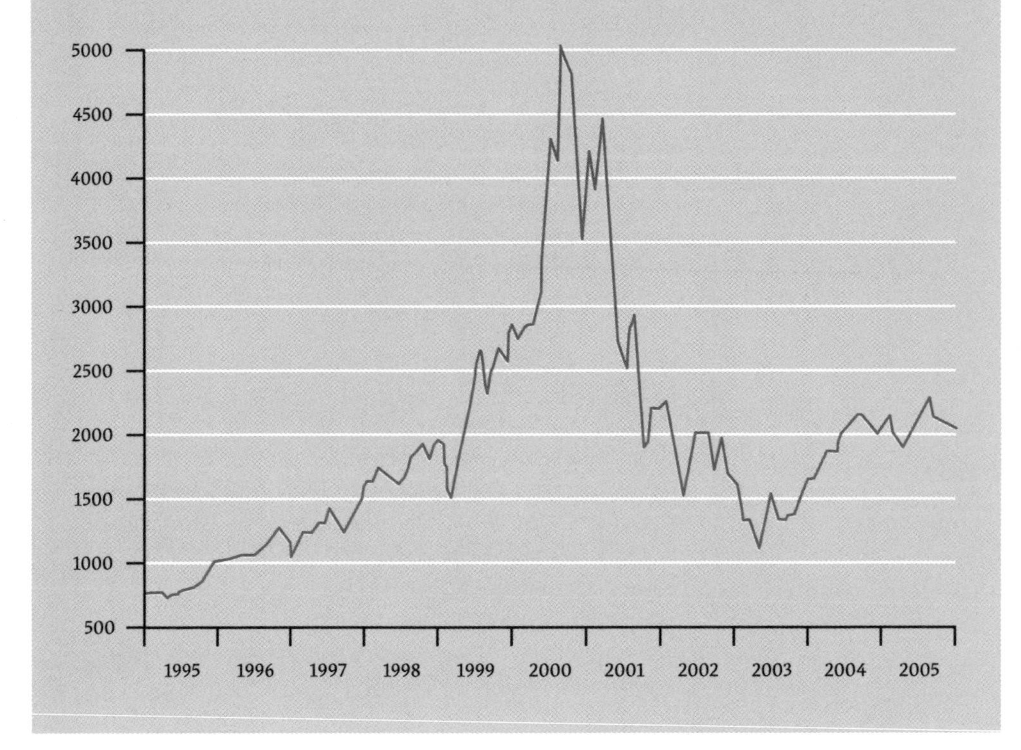

Bear in mind this is a classic example of chaotic behavior: There is clear self-similar structure in the data, but knowing what that structure is gives us no help in predicting what will happen next. And the power that needs to be applied could easily change from one episode to the next with the nature of dominant businesses. Interestingly, this power law distribution is rather similar to a natural collection of shocks, in earthquakes—though in that case, halving the size typically doubles the number of events rather than tripling them—and for that matter is reminiscent of the widescale applicability of the Feigenbaum number.

From pundits to bestsellers

"A best-seller is the gilded tomb of a mediocre talent."
Logan Pearsall Smith, 1865–1946

Political confusion

We have already seen that polls (and bookies) seem to have been struggling with predicting the outcomes of political decisions in recent years. In a broader sense, the political pundit has faced increasing difficulty with a system that has become increasingly chaotic. A large factor in this is the move away from voting occurring on traditional socioeconomic and party lines. Divisions have been changing and fracturing, producing the ideal situation for chaos to arise from the interplay of unexpected parts of the system.

There seems little doubt that a familiar factor here is the growth of social media and other online forms of communication. Where traditionally political influences might have been limited to where we grew up, our workplace, and the mass communications of political parties, now there are far more opportunities for information to circulate within social bubbles, reinforcing what might previously have been relatively minor factors. The result has been outcomes that have taken the pundits and often the "metropolitan elite" by surprise, as what was once a strongly structured system where traditional probabilities played out has become a chaotic one where outcomes that would previously have seemed impossible come true.

Going for gold

Unlike the political environment, the system for what makes a bestselling book, or a number one song, has always had a significant chaotic component, though the same factors, such as social media and other online environments (notably Amazon), have only made this chaotic aspect more central. If we take bestselling books as an example, this is arguably a far less understood chaotic system than one that has been well studied, such as the weather.

Although there are clear factors that will contribute to the success of a book—the marketing and visibility it receives, media and social media coverage, disposable income of the target audience, changing attitudes to reading, and more—this is a complex system involving the actions of publishers, bookstores, and buyers where the data is far more sparse than is the case with the weather, and where the available relationships are far less obvious than, say, the gravitational interaction of the three-body problem (see page 59).

The outcome has been made more complex by the availability of e-books, which allow for a much greater flexibility of pricing, plus the interactions within large new systems, such as an online retail environment like Amazon. Say there is an e-book priced at $4.99, which is temporarily reduced in price to 99 cents. Conventional economics would allow a prediction to be made of the impact such a price drop would have on sales. But in the Amazon ecosystem, the price change is only a small part of the interacting factors. Because of the drop, the book may appear in a special section for deals. The increase in sales could push the ebook up the sales ranking, giving it a lot more visibility on the site, further increasing sales in a positive feedback loop. Suddenly what seems like a relatively minor change could be hugely significant—but in a totally unpredictable fashion.

This is not saying a book of absolutely any quality can become a bestseller (though there are examples, such as *Fifty Shades of Grey,* which have had a huge success despite being totally panned critically), rather, that it is almost impossible to predict what will be a bestseller, because there is no clear, simple relationship between any particular factor and large-scale success. This is despite attempts that have been made to use big data to make exactly this kind of prediction.

Big data

"Mathematics may be compared to a mill of exquisite workmanship, which grinds you stuff of any degree of fineness; but, nevertheless, what you get out depends on what you put in; and as the grandest mill in the world will not extract wheat-flour from peascods, so pages of formulae will not get a definite result out of loose data."
Thomas Huxley, 1825–1895

Is big really beautiful?
We live in an age of big data, having unrivalled ability to take vast amounts of data, often nearly in real time, and analyze it. Some would say that this is the solution to the problems of polls and surveys—forget trying to work from a sample of the whole when you are dealing with chaotic systems, and simply pile in all the data. Arguably the first attempt at big data, in the form of a census, is what triggered the computer revolution. A census is an attempt to take a snapshot of everyone in the system being studied, but by the time of the 1890 U.S. census, it was taking so long to process all the data that there was a danger that the task would not be completed before the 1900 census took place.

To the rescue came the Tabulating Machine Company, using a punched card system developed by American inventor Herman Hollerith. His devices were not computers—they simply sorted and selected cards electromechanically—but his company would become IBM and his card-based data processing systems were the predecessors of their computer-based equivalents.

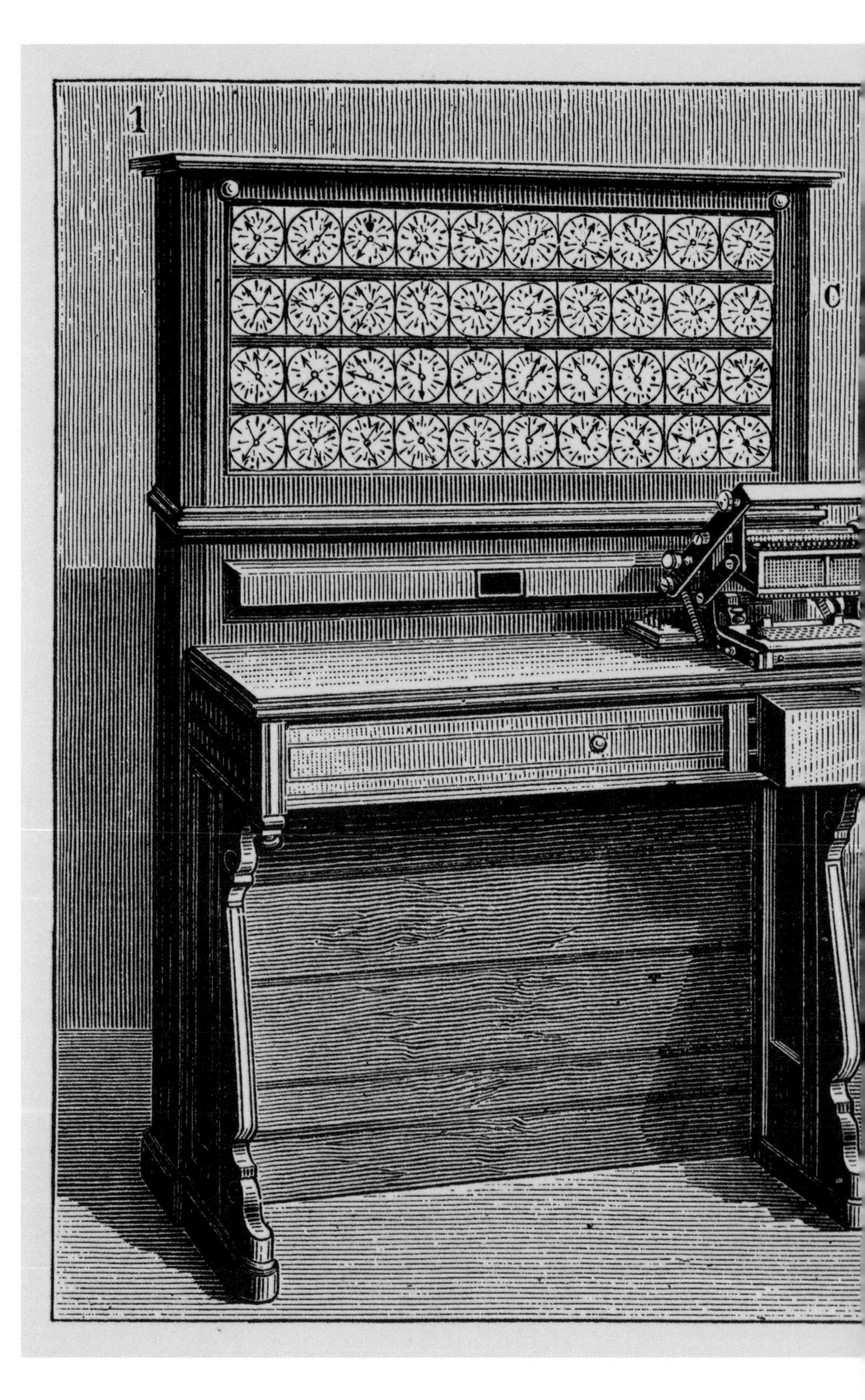
1
C

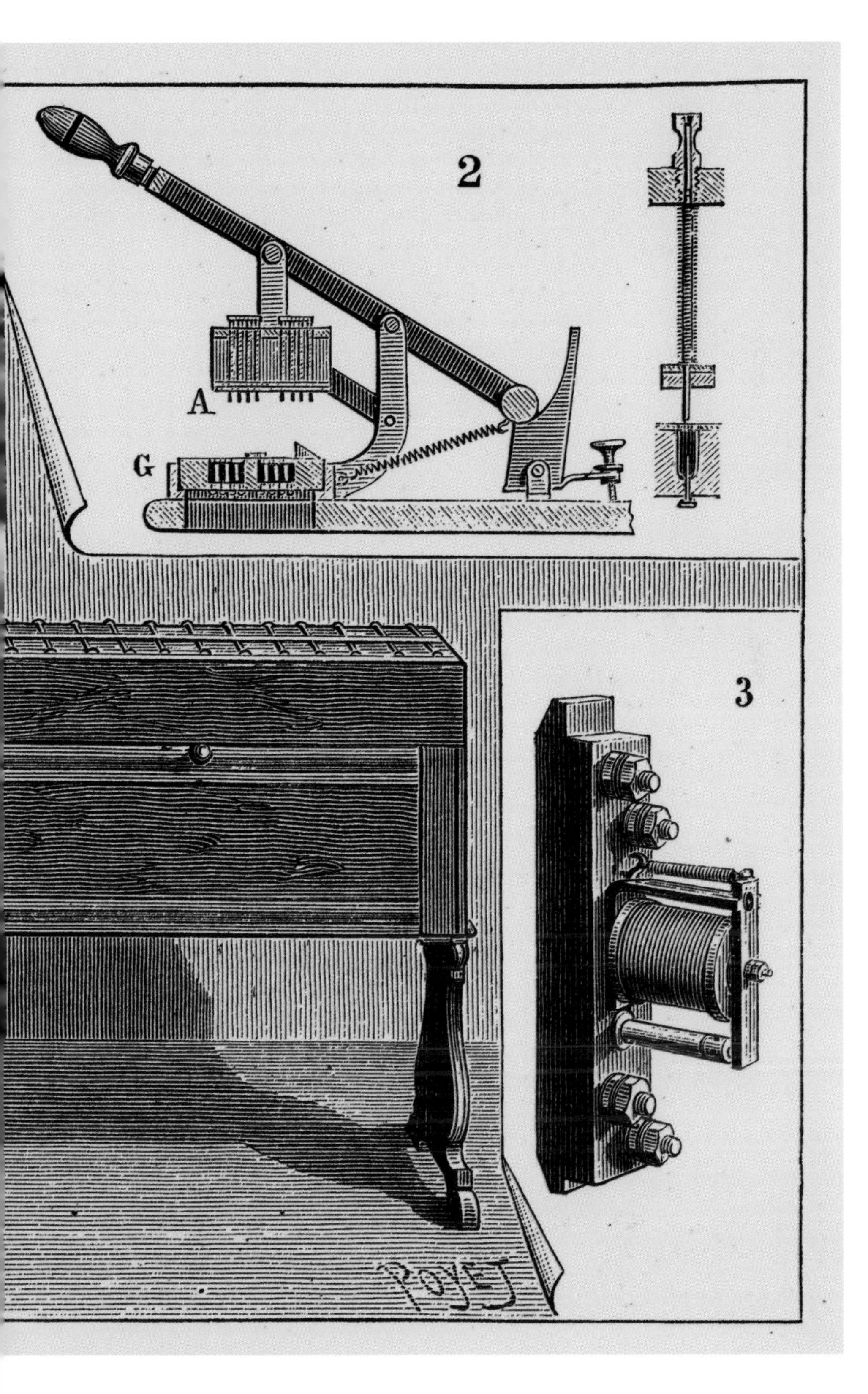

Hollerith tabulator
An 1890 tabulator as used in the U.S. census.

However, a census is a one-off measure. Big data in the modern sense would not take off until the twenty-first century, when companies and organizations began to collect and process large quantities of data about people and their actions—and when software started to become clever enough to begin to sift through and manipulate that data without human intervention. Often this incorporates elements of artificial intelligence (AI) or "machine learning," where, rather than humans setting the rules for working on the data, the computer discovers links and associations on its own.

In principle this is a boon. We know that we struggle to understand chaotic systems—which are almost always what we are dealing with when people are involved. (This sounds sarcastic, but it is not because people's lives are chaotic but because there are a lot of interacting factors shaping what they do.) So why not let an artificial intelligence system get on top of what's happening and make predictions for us? This has been tried for everything from credit scoring to predicting where crimes will be committed. But there are serious issues in taking this approach.

Recognition requires understanding

Two classic problems with allowing AI systems to trawl a vast pool of data and make deductions and predictions for us are that, first, they have no understanding and, second, they show a tendency to overfit. Let's look at each in turn. It's easy to be fooled by the term "artificial intelligence." AI systems are *not* intelligent. They find patterns and exploit them—which certainly is a tool used by intelligence—but the AI systems have no understanding of what they are doing.

A good example comes from the attempt to use AI in image recognition. Identifying what is in a picture is something humans are extremely good at—but, traditionally, machines are not. AI seems to offer a similar capability to the human. Let's say we want image recognition software to pick out images featuring a pair of skis. It's very difficult to specify exactly what a pair of skis will look like, seen at any angle from any direction. But machine learning systems don't need us to do this. Instead we show them millions of photographs and tell them which ones have skis in them. Over time, the system will get better and better at recognizing skis. Or so it seems.

→
Self-driving car
The self-driving car's picture of the road ahead depends on reliably recognizing road signs.

The catch is that we don't know what the system is using to identify skis—it certainly doesn't recognize what skis *are* in the way that we do. It's just that it does pick out the right images well across the photographs we've shown it. But then we start to use the system, and whenever we show it a photograph with a snowy background, it tells us there are skis in the shot. Because the vast majority of photos of skis the machine was shown when learning also had snow in them—and a wide field of snow is a lot easier to spot than a pair of skis—it was really looking for snow all along.

Even worse, programmers have been able to demonstrate that it is always possible to fool an AI image-recognition system using something that, to the human eye, is negligible. A worrying example comes from the recognition systems used in self-driving cars. It is possible to produce a small sticker, which to us is nothing more than a scrambled mess; but when it is attached to a stop sign, a self-driving car reads it as a speed limit sign. Using these systems to recognize something that is then checked by a human is not a problem, but to give such a system the ability to act based purely on its own decisions is worrying.

We've already seen how easy it is to come up with correlations that make no sense, such as Vigen's oil imports and collisions with trains example on page 158. It's possible to do this because individuals seeking spurious correlations have access to large amounts of data to trawl through. With enough data, there are bound to be some coincidental correlations. But trawling through large amounts of data is exactly what big-data AI systems do. And because machines lack understanding, unlike us, they can't spot that the apparent link they have discovered is silly. And then there's the matter of overfitting.

Overfitting reality

Forecasters have a problem. What they usually do is to take some past data and try to extrapolate this into the future. The argument is: *This is what happened before, I should be able to deduce* (strictly, to induce) *from it what comes next.* Unfortunately, reality is rarely nice and smooth in its behavior. Even non-chaotic data will have "outliers"—examples where something went temporarily away from the norm—while, as we have seen, chaotic data will have far more extreme jumps. It is part of a chaotic system's very nature.

When trying to work out the shape of what has happened to project it into the future, it has long been recognized that there is a danger of "overfitting"—of taking every single point of data and producing a model that would be able to generate exactly that set of data. The trouble is that the model has then become so precisely pinned down that it can only keep reproducing the same outcome. In practice, a model needs to have a looser fit to the data if it is able to take on the future, typically just using a relatively small number of variables.

Left to its own devices, though, a machine learning system can easily wildly overfit the data, making up vast numbers of

Overfitting
A sensible fit for the border between black and white dots is the black line. Fitting the line to every single dot (red line) produces a plot that only works with this very specific set of data.

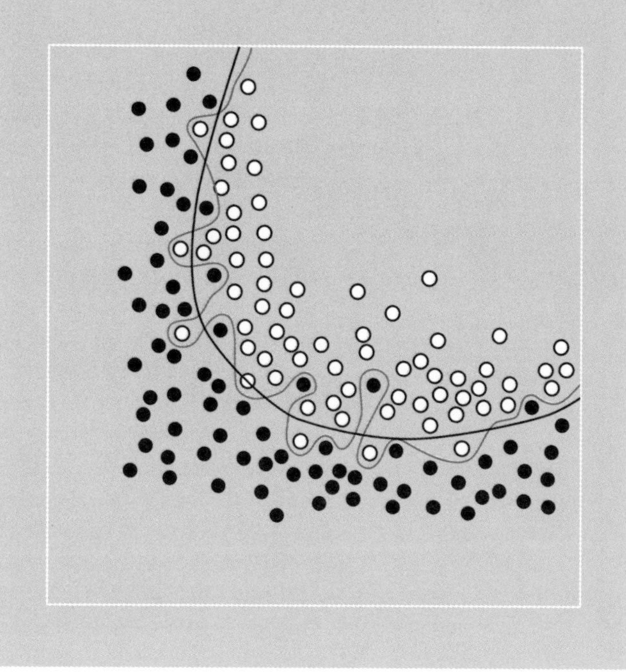

parameters that have no basis in reality but nonetheless enable it to predict exactly how its training data—the data it learned from—behaved. This tends to make such a system useless for any application away from the data it was trained on. Of course, developers of such systems attempt to prevent this, but the whole point of machine learning is that we don't tell the systems how to go about their job. And once again, the lack of understanding is an issue.

Bestsellers revisited

As we have seen, publishers have no good way to determine how the chaotic system that is the book-buying public will respond to a new title. Yet in a book called *The Bestseller Code: Anatomy of the Blockbuster Novel*, American academic Matthew Jockers and editor Jodie Archer suggested it was possible to use AI and big data to predict the potential success of books. The authors said that their system "can read, recognize, and sift through thousands of features in thousands of books"—and that's a recipe for generating coincidental correlations and for overfitting.

The authors make no claim that their system will pick out good literature—it's all about the potential sales. And there is no doubt that it would identify some "me too" books that could do well on a bestseller list. But it's clear from the list of topics the Jockers-Archer model suggests should be avoided by an author that the approach has serious problems. To be a bestseller, according to this system, an author should avoid, for example, fantasy, very British topics, sex, and description of bodies. So that means, to pick at random, *Game of Thrones* and *Lord of the Rings* aren't potential bestsellers, not to mention characters such as James Bond or Harry Potter, the majority of successful young adult titles, any murder mysteries, and *Fifty Shades of Grey*. The approach may generate relatively safe titles—but following such rules certainly isn't a way to produce great original fiction or the next new big thing. That isn't how chaos works.

Overall, then, there are serious concerns about AI systems that can influence our lives, basing decisions on factors that may make no sense to anyone with understanding. It's not surprising that many governments are wary and are looking for ways to protect the public by requiring those using such systems to build in transparency. But that's easier said than done.

Transparency
Because a machine-learning AI system derives its own rules from a mass of data, it can be extremely difficult to provide a logical explanation for a decision. For an AI system to be transparent it needs to formulate a logical process that can be followed by a human being.

When a machine learning system makes a decision, the factors on which it bases that decision can often be messy and will have no obvious logic behind them. It's arguable that requiring transparency and explanations of decisions will mean either abandoning the AI systems or, more likely, building a lot more complexity into them to enable them to construct a logical reason for a decision, rather than a set of incomprehensible weightings for different factors. Those working in the industry sometimes complain that the legislators don't understand the technology—but in doing so, the technologists underline the problem that we all face. Unless the technology can do the job including providing explanations, it is the technology that is failing in its role.

Emerging from chaos

"Science is the attempt to make the chaotic diversity of our sense-experience correspond to a logically uniform system of thought."
Albert Einstein, 1879–1955

Chaos isn't always chaotic
Before the word "chaos" was given its mathematical definition in the 1970s, it only had the original meaning used by Einstein in the quote above: a random, often dangerous, formless mess. It really would have been better if those who applied the term to its mathematical definition had come up with a different word, because it's very difficult to separate the term from the underlying strength of that original meaning. Quite rightly, we expect chaos to be an anarchic mess. And it can be. But much of the time it isn't.

Think for a moment of the original discovery by Lorenz. The weather certainly can be chaotic in the traditional sense. But often it isn't. Imagine serene blue skies or even a steady fall of rain from a blanket of cloud. The weather is still mathematically chaotic—but it does not necessarily bring with it the attributes of chaos in its conventional English-language sense or as the primordial void in Greek mythology.

We have already seen that chaotic systems often have attractors—islands of calm that are relatively easy to reach from a wide range of starting points. The route to get there may be entirely unpredictable and crazily variable, but the destination is still anything but disordered. Equally, it is possible for chaotic systems to result in emergence into a surprising degree of something that might be regarded as the antithesis of chaos: synchronicity.

Spontaneous synchronicity

Satisfyingly, like chaos, synchronicity is another word that doesn't quite mean in a technical sense what it is generally used to mean. It feels (it's not exactly in common usage) as if it means a state of being synchronized. So a system that exhibits synchronicity has a regularity of something time dependent, say motion, that stretches across part or all of the system. In reality, the term was devised by Swiss psychologist Carl Jung to describe events that coincide in time and appear to be related, but with no obvious causality. (That's a coincidence, Carl.)

In this definition, then, synchronicity is about correlation without causality. Jung gave the example of the Chinese fortune-telling system, the *I Ching*. This will sometimes seem to give meaningful predictions as a result of coincidences, but there is no causal mechanism. Chaos can result in a different version of synchronicity that makes more sense, almost an inversion of Jung's, where there *is* a causal link, but it isn't obvious, thanks to the chaotic nature of the system.

One of the first chaotic systems we met was a hinged pendulum (see page 38). An alternative pendulum, a so-called coupled pendulum, shows how chaos can produce synchronized action. Here two pendulums of the same length are suspended from the same object. When one pendulum is started, the oscillation will gradually transfer to the other pendulum, then back to the first one and so on. When both are moving, their movements synchronize. This was first observed by the Dutch scientist behind the pendulum clock, Christiaan Huygens.

This wasn't an arbitrary discovery. In the seventeenth century, ocean navigators had a major problem in determining the longitude of their vessel—the east–west position on Earth's surface. Latitude is relatively easily measured from the Sun's position at midday, but determining longitude required an accurate measure of the time difference between the current location and a known point on Earth. Teamed up with Scottish scientist Alexander Bruce, Huygens wanted to see if his newly invented pendulum clocks could cope with a ship's motion at sea, and arranged to have a pair of pendulum clocks installed on a ship heading out from west Africa.

→
The *I Ching*
A Song Dynasty (960–1279) page from a printed copy of the *I Ching*.

纂圖互註周易上經乾傳第一

王弼註

乾下乾上 ☰ 乾元亨利貞初九潛龍勿用

九二見龍在田利見大人

九三君子終日乾乾夕惕若厲无咎

Synchronizing pendulums
In Huygens's experiment, the pendulums of two clocks fixed to the same wooden beam would synchronize in opposition.

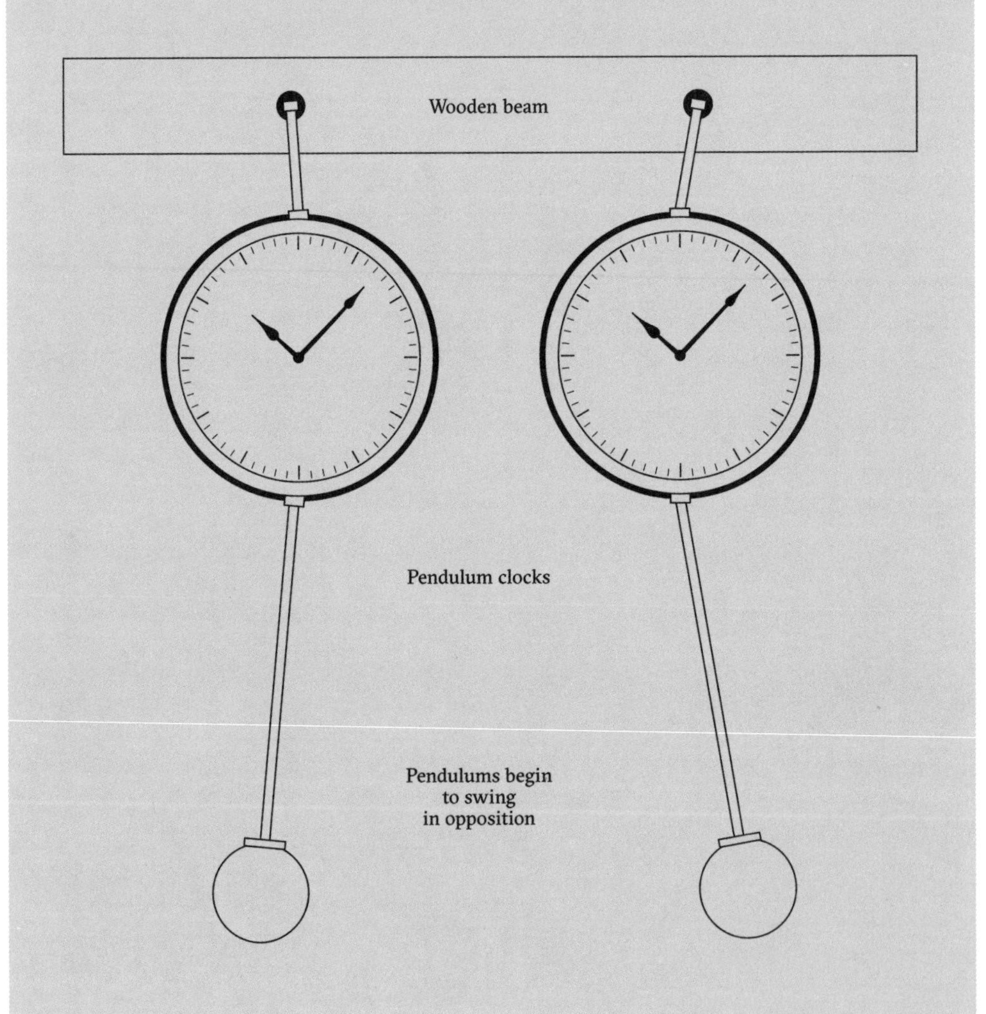

There were two clocks to provide a backup—if one happened to be stopped by the violent motion of the seas, it could easily be reset to the time given by the other. As it turned out, although no storms were encountered and the initial trial was successful, the approach proved ineffective. This was in part because the clocks weren't accurate enough and also because they were too susceptible to the ship's motion, making it entirely possible both would stop at once. But in trying different ways to mount them, Huygens discovered that two clocks fixed to the same wooden beam would synchronize, swinging in time with each other, though in opposite directions, after about a half hour in action.

This effect was rediscovered a number of times over the years, and by the nineteenth century there was a clear explanation that energy was being transferred between the pendulums through the bar that connected them, initially chaotic until they were exactly out of phase with each other—mirroring each other's actions—then the energy transfers between the pendulums balanced each other out.

More recent research has uncovered further remarkable aspects to pendulum synchronization. In 2002, James Pantaleone, a physicist working at the University of Alaska, tried an experimental setup using two metronomes—those inverted pendulums used to provide a regular ticking sound to set times in music practice—sitting on a wooden board on top of a pair of soda cans, in what should have been a cheap and cheerful demonstration of the effect discovered by Huygens. Instead, something strange happened.

The metronomes did synchronize—but instead of settling down and moving as mirror images of each other, they synchronized, both moving in the same direction. Further experiments at Eindhoven University of Technology in the Netherlands showed that the secret was not the drink cans, but the relative lightness of the linking bar. With a lightweight bar, the pendulums synchronize in the same direction, with a heavy bar in opposite directions. It was established that, rather than batting energy back and forth, with the light bar the motions were effectively coupling through the bar, as if the pendulums were connected to each other.

Moons, fireflies, and bridges

A similar effect produces a phenomenon that is so familiar that we rarely think about it, but that can seem puzzling when noticed. The Moon always shows the same face to us. It doesn't have a "dark side" as suggested, for instance, by Pink Floyd—the far side of the Moon receives just as much sunlight as the near side. But the same side always faces toward Earth. As the Moon is orbiting around our planet, the only explanation for this is that the Moon's rotation speed is just the right amount to cancel out the changes we would expect to see as it passes around us.

This seems a remarkable coincidence—and it would indeed be remarkable, if it were just a coincidence. However, it's another example of synchronicity. The Moon and Earth are not perfect spheres. Most importantly in this respect, because Earth's gravitational pull is significantly stronger on the side of the Moon facing Earth, that side is slightly stretched out toward Earth. This acts a little like having a weight on one side of die. When a weighted die is rolled it is more likely to come up with the value on the opposite side to the weight. Similarly, the side of the Moon that bulges toward Earth feels more of a pull toward Earth as the Moon rotates. Over time, the Moon's rotation would have been nudged by this pull toward a speed that kept one side always facing toward our planet.

Another example of synchronicity in nature is that some species of firefly will synchronize their flashes, seemingly not through conscious intent to do so, but because of their interaction in a chaotic system.

One of the best examples of such spontaneous synchronicity occurred when London's Millennium bridge was opened. This is a steel suspension footbridge crossing the River Thames between St. Paul's Cathedral and the Tate Modern art gallery, and was built as its name suggests to mark the new millennium. When first opened, the bridge was, well, a walking disaster. Despite being a large, solid structure, it bounced around as people walked over it, so much so that it was difficult to cross. Within hours of opening, it had to be closed.

The bridge's problem was fixed by attaching dampers to stop the behavior. But the cause of its bouncing was the emergence of synchronous motion from the chaotic input of so many pedestrians. This is different from the old idea of soldiers having to break step when crossing a bridge in case their marching

Dampers
Mechanical devices that absorb sudden motions, often by the use of springs or hydraulic cylinders. Shock absorbers on cars are a form of damper.

fitted with the frequency at which the bridge naturally vibrated. Here the motion imparted to the bridge was up and down. These people weren't walking in step. But as we walk along, we naturally sway from side to side. From the chaos of a whole host of people crossing, synchronicity emerged: if there were slightly more people swaying one way than the other it caused a small amount of side-to-side motion of the bridge deck. This encouraged other people unconsciously to sway in time with it, amplifying the effect.

Chaos, then, can defy expectations. And we have already seen what a problem it presents for those attempting to predict the behavior of chaotic systems using the traditional mathematics of randomness. So how can chaos theory give us a practical advantage? It is one thing to be aware of chaos and another to harness it.

Synchronous fireflies
Fireflies in Smoky Mountains National Park, Tennessee, flash in synchronicity.

6
Harnessing Chaos

Turbulent times

"When I meet God, I am going to ask him two questions: Why relativity? And why turbulence? I really believe he will have an answer for the first."
Werner Heisenberg, 1901–1976
(also attributed to Horace Lamb, 1849–1934)

The frequent flyer's roller coaster
It's a moment that few air travelers relish—when the captain's voice comes over the intercom to announce that the plane is about to hit turbulence and everyone needs to fasten their seatbelts. It's one thing to be seven miles high in the sky cruising smoothly and another to be up there bouncing around as the plane rattles and shakes like a toy in the hands of a child. It's even worse when the plane hits a pocket of clear air turbulence, as a result of which the aircraft suddenly drops like an out-of-control elevator.

Something that should be reassuring if you ever encounter turbulence on a flight—no large passenger plane has ever been brought down by turbulence (small planes have). But plenty of people have been injured, either by items falling out of lockers or, in extreme cases, by not being strapped in and flying out of their seat. But arguably what should be the most unnerving thing about experiencing turbulence is that it is one of the least well-understood aspects of physics. For a long time, physicists found this deeply frustrating, but now they realize it is because turbulence is a characteristic of a chaotic fluid system.

Air is a fluid, and the turbulence an airliner experiences is caused by sudden movements of the air, typically when adjacent regions are at significantly different temperatures, producing strong flows of air from one place to another. The movement of fluids is described by a set of equations known as the Navier–Stokes equations after French engineer Claude-Louis Navier and Irish-born English physicist George Stokes. The joint naming does not indicate that the two worked together, but resulted from an unusual combination of contributions.

Navier came up with the equations in 1822 (when Stokes was age three), but he seems to have done so in error. Navier did not understand the physics involved and was unaware that he needed to deal with what is effectively friction between the fluid atoms or molecules—but his equations did, somehow, manage to correctly represent this. Stokes's name is appended because in 1845, after Navier's death, he correctly worked out the physics and derived the same equations, putting them on a solid scientific footing.

In the most general case, the equations are what are known as nonlinear partial differential equations: such equations are complex enough to make them only fully soluble for relatively simple cases. The "nonlinear" part is a strong hint that chaos will follow, referring as it does to mathematics where the output is not simply related to the input. Turbulence arises when different parts of the fluid accelerate at different rates at different points in time, so that the fluid does not move as a smooth whole or in a set of layers (known as laminar flow), but whereas small parts of the fluid have unpredictable motion in different directions.

Heart eddies

Sadly, and this is becoming a common theme, although an understanding of chaos helps us explain why turbulence occurs, it does not provide us with the tools to predict its behavior any better. There are models of turbulent flow that

→

Wingtip vortices
As the plane's wingtips cut through the air they cause turbulence, made visible by the contrails where the drop in air pressure causes water vapor to condense.

modify Navier–Stokes equations (for example, by averaging over time to reduce the impact of time-based variations in acceleration) or that model fluid behavior numerically, but they cannot address the chaotic outcome of a specific example of turbulence. Even so, they help us to understand what is happening, which is valuable, for example, when turbulence occurs in the heart.

So-called heart murmurs are usually caused by turbulence of the blood flow within the heart, for example when a heart valve is slightly deformed, meaning that the blood flow around it is no longer smooth and the result is eddies and currents that can be felt by the patient or heard externally.

Wingtip wonders

We return to aircraft for a third example of turbulence. It's not uncommon for a plane to get to the end of the runway ready for takeoff, but then to wait a minute or two for no obvious reason, to the frustration of the passengers. The plane in front has already taken off, but still they have to pause. If another plane has just taken off, it is likely to have left behind a form of turbulence that could make takeoff dangerously unstable for the next aircraft, so the plane then has to wait.

This turbulence is caused by wingtips cutting through the air. Once the plane is moving at speed, the relatively narrow wingtips exert strong pressure on a localized section of the air ahead of them, producing powerful sheering forces, which generate local turbulence that exhibits itself as decaying vortices, spinning spirals of air trailing from the wingtips. As a result, modern planes tend to have technology built into the wing to minimize these effects.

In some cases, turbulence is reduced by using winglets—small extensions sticking upward from the wingtips. ("Small" is a relative term here—the winglets on a large airliner are taller than a person.) A winglet cuts through the vortex, breaking it up before it can properly form. This isn't specifically to help planes waiting in line on the runway, but rather because the vortices add drag to the wing (and as a bonus, the winglet increases lift). Large modern planes without winglets have special shaping on the outer section of the wing to disrupt a vortex.

For wingtip turbulence, an understanding of chaos helps us with something we want flying through the air—but there is also a more sinister possibility where the study of chaos may help.

Winglet turbulence reduction
A comparison of the interaction of aircraft wingtips with the air, without and with winglets. The winglet (shown on the right) significantly reduces the ability to generate vortices.

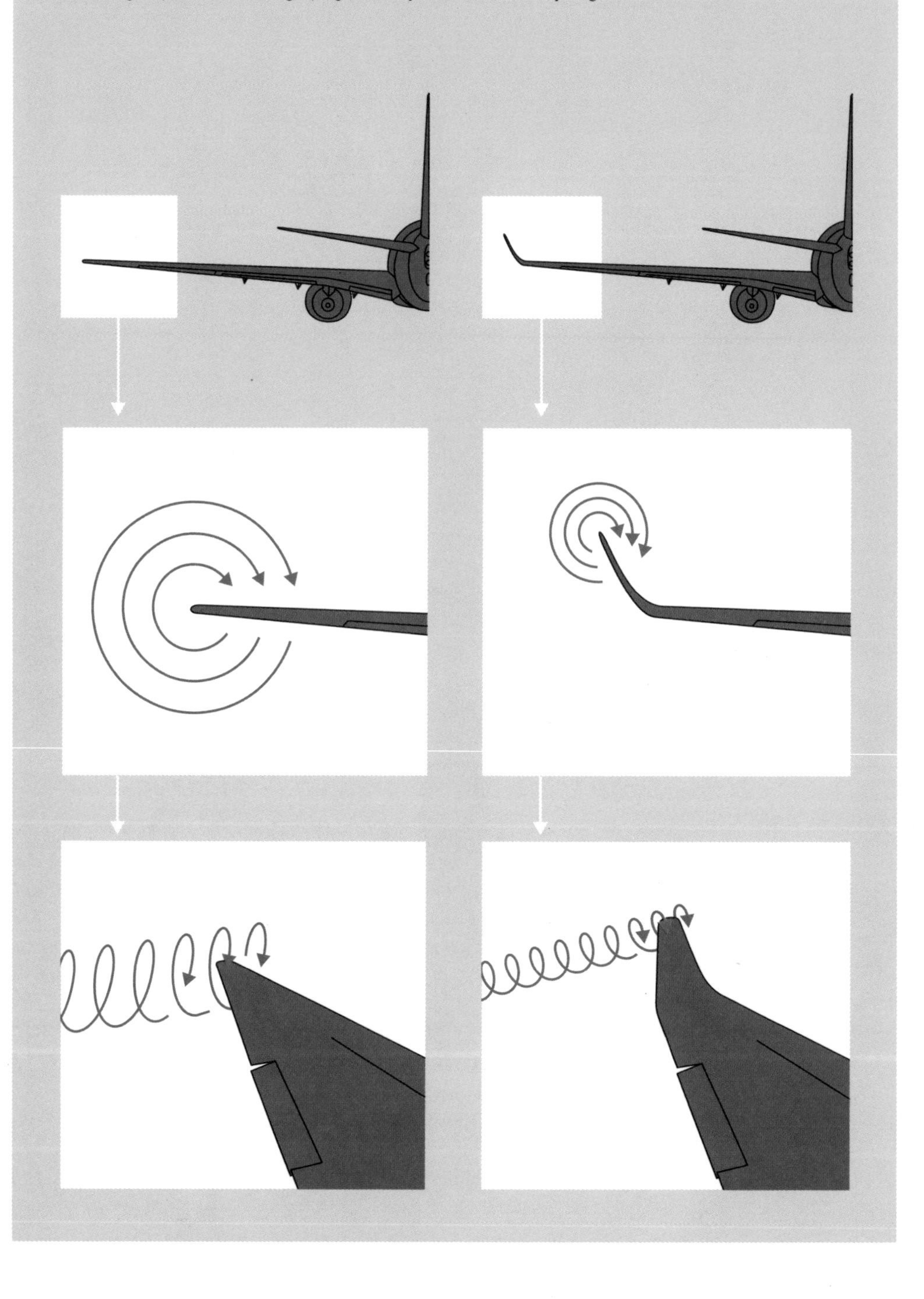

Danger from the sky

"Now it must be asked if we can comprehend why comets signify the death of magnates and coming wars, for writers of philosophy say so."
Albertus Magnus, ca. 1200–1280

Death to the dinosaurs
The very fact that humans exist is (in part) down to a catastrophic event around 65 million years ago, an event that was determined by chaos. The Yucatán peninsula on the Gulf of Mexico was hit by a 6-mile (10-kilometer) wide (estimated sizes vary) chunk of rocky material traveling at around 12.5 miles (20 kilometers) per second. The amount of energy released in the impact was vast—around five billion times the energy produced by the atomic bomb that was dropped on Hiroshima during World War II. For hundreds of miles around the impact site everything would have been killed—but, more significantly, the vast cloud of ash and dust thrown up must have darkened the atmosphere for years, cooling the planet and killing off the dinosaurs, making it possible for our more adaptable mammalian ancestors to thrive.

This was not the first time Earth has been hit by an asteroid or comet—nor was it the last, though there hasn't been an impact on that scale since. On Earth it's difficult to see where such impacts have occurred, because tectonic activity, the oceans, and living organisms all tend to disturb the surface and conceal the remnants of the collision. The resultant Chicxulub crater on the Gulf of Mexico, around 125 miles across (200 kilometers),

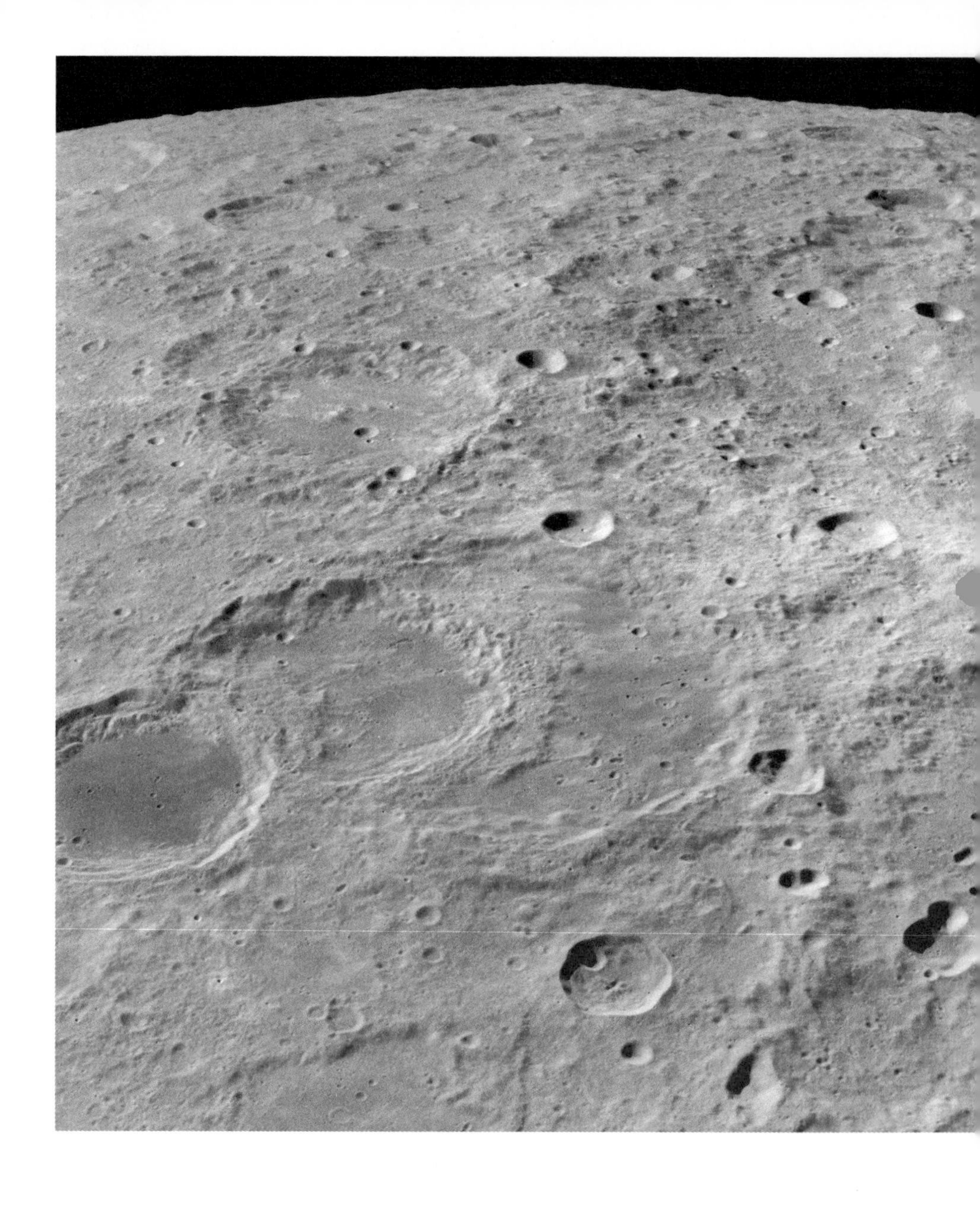

The scarred surface of the Moon
Without tectonic activity, oceans, and the impact of living organisms, the Moon carries the scars of its impact craters in a far more visible fashion than Earth.

was not spotted until 1978, during oil prospecting, as it is largely concealed. But look at the Moon, where such impacts tend to remain visible, and you can see the results of billions of years of pounding from space.

We have had relatively small extra-terrestrial collisions in recorded history. Perhaps most famous is the impactor, possibly part of a comet, that exploded over the Tunguska river valley in Russia in 1908. Thankfully, this was a largely uninhabited region, but the object was over 150 feet (45 meters) across and flattened trees around it in a 19-mile (30-kilometer) radius. The most spectacular example that occurred in modern memory was also in Russia—the meteor that exploded near to Chelyabinsk in 2013. This was significantly smaller and exploded well up in the atmosphere but still produced the energy of a sizable nuclear weapon.

Tunguska event
Air waves from the explosion near the Podkamennaya Tunguska river in Siberia were detected in Washington D.C. It has been blamed in fiction on everything from visiting aliens to a miniature black hole.

We can't say when, but there will be another catastrophic collision at some time in the future.

Incoming alert

The dinosaurs had no opportunity to prepare for their impactor. But with our understanding of physics and far-ranging telescopes, we do. However, chaos is ready to ensure that we don't get an easy ride. As we've already seen, just three bodies interacting makes for a gravitational problem that we can't precisely solve. When a potential collider is spotted heading in our direction, we can plot its future path—but only so far. For example, an asteroid that is apparently in a safe orbit in the belt between Mars and Jupiter is influenced by the other asteroids around it and the Sun—but also by the looming presence of Jupiter. The giant planet's gravitational pull can make gradual changes to an asteroid's orbit, which can eventually send it off in a totally different direction. And in attempting to predict the future result of those changes, we hit against the difficulties imposed by chaos.

Things are even worse with comets, which don't have simple planetlike orbits keeping them to a particular region of the solar system. Instead they plunge in toward the Sun from way outside the main planets, potentially passing near a whole range of influential massive bodies, which can make drastic changes to their elongated orbits. To make matters even more chaotic, comets are able to generate their own rocket motors. As they are heated by the Sun, water and other volatile

Asteroids in the solar system
The solar system's asteroids are largely found in two locations: the belt between Mars and Jupiter and the so-called Trojan asteroids, held in place around Jupiter's orbit by the interaction between the Sun and the planet's gravitational fields.

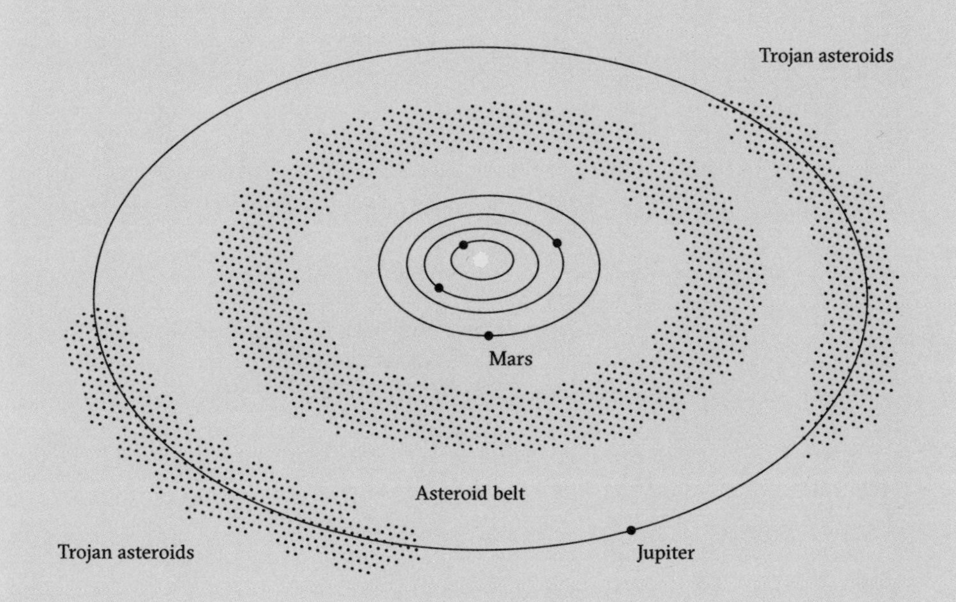

→

Potentially hazardous asteroids
The orbits of the over 1,400 asteroids identified by 2013 that are at least 460 feet (140 meters) across and have orbits that pass within 4.7 million miles (7.5 million kilometers) of Earth.

materials can evaporate from them, spraying out a jet of gas and dust which acts just like a rocket motor, pushing the comet off course.

Of course, we are able to predict the orbits of recurring comets well ahead of time in most cases. Famously, Newton's contemporary Edmond Halley successfully predicted the return of the comet named after him, even though it did not become visible again until after his death. But other comets can, after many relatively stable orbits, move just a bit too far in one direction and suddenly be pulled off course.

As always with the mathematics of chaos, knowing that a system is chaotic does not magically give us a mechanism to solve the equations and give us a perfect prediction of what will happen. However, the same kind of ensemble forecasting approach used to such good effect to improve weather forecasts can also be used to attach probabilities to different trajectories of potentially dangerous asteroids and comets. At least, that is the case if we know there is a chance of an impactor heading too close to Earth's orbit.

Spaceguard

Thankfully, we do have a loose group of organizations and observatories, working under the title of Spaceguard (a clear case of life imitating art, as the name was taken from the asteroid collision protection organization in Arthur C. Clarke's 1973 science-fiction novel *Rendezvous with Rama*). Spaceguard is an international effort, spearheaded by the United States, that watches the skies for incoming bodies that could become "near-Earth objects"—those that are close enough to Earth to present a significant risk of collision.

It was one of the systems that form part of Spaceguard, the Panoramic Survey Telescope and Rapid Response System, based on Maui in Hawaii, that spotted the strange cigar-shape object 'Oumuamua in 2017 that would go on to pass around the Sun, though this never came close to an orbit that risked collision. Because of its unusual shape, it was speculated that 'Oumuamua could be some kind of spaceship, but nothing in its behavior, tracked throughout its pass, suggested it was anything other than space debris.

Although there can't be total certainty of an incoming object's trajectory when it is first observed, the idea of Spaceguard

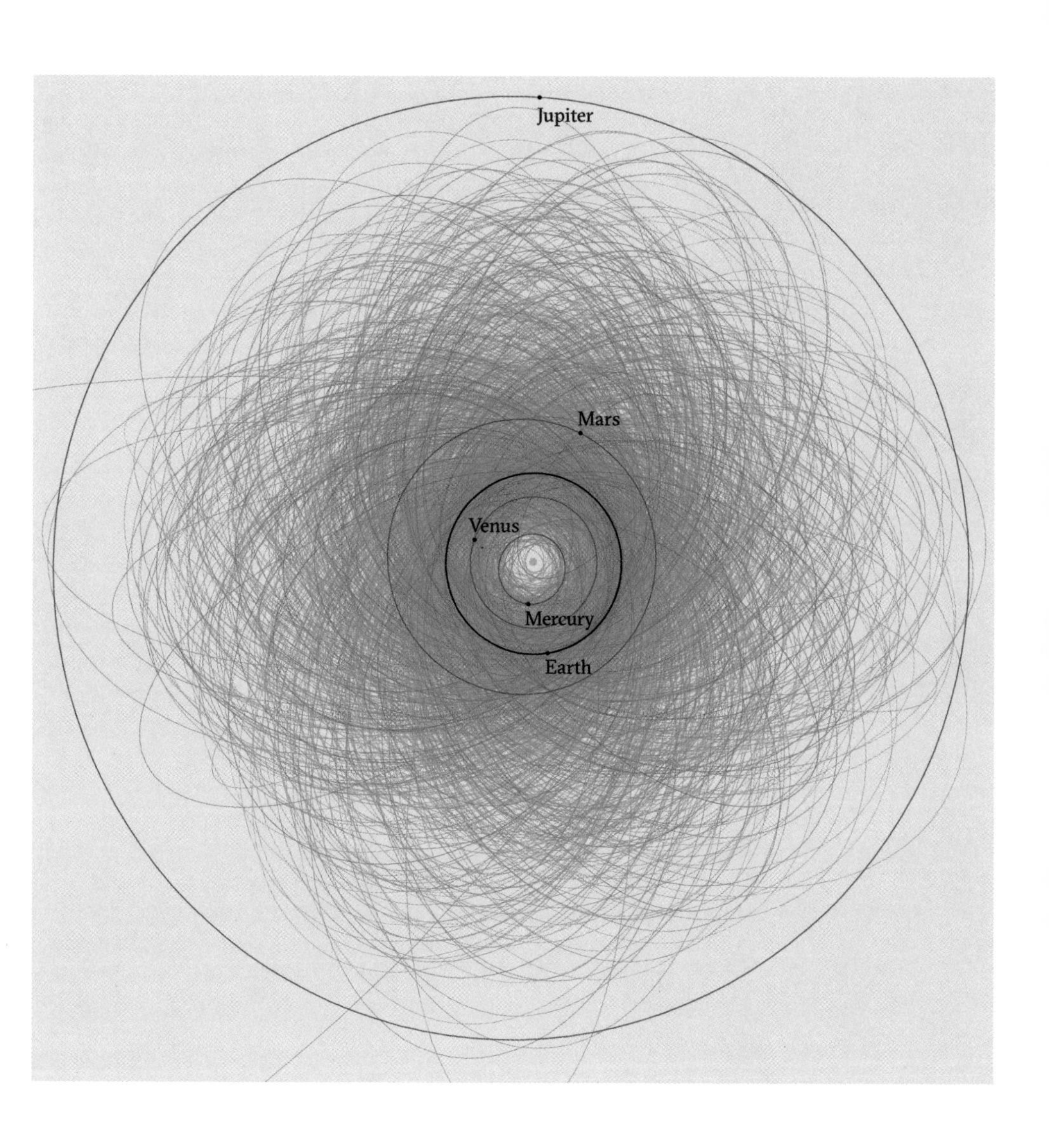
Jupiter
Mars
Venus
Mercury
Earth

is that all nations would be alerted early enough to consider some sort of intervention if the risk levels become high enough to warrant it. This could involve hitting the object with a high-speed probe, placing motors on the incoming body to slightly change its path, or even using bombs to shatter it, though this last is a particularly risky approach because unless carefully managed, the debris from the exploded asteroid or comet could still hit Earth and cause extensive damage. NASA is planning a test of an asteroid interception spacecraft called DART, which is intended to use impact to change the orbit of a near-Earth asteroid called Didymos.

A natural assumption might be that the purpose of an attempt to change the object's path through space would be to move it sideways on its course, but in fact there is usually less of a change required if the potential collider is either slowed down or speeded up. Bear in mind that Earth is not a stationary target. Our home planet is moving at high speed through space—at around 67,000 miles per hour (107,000 kilometers per hour) with respect to the Sun. If an incoming object is on a collision course with Earth but arrives a little late or too early, Earth will no longer be in its path by the time it arrives.

While chaos gets in the way of Spaceguard being 100 percent certain of keeping us safe, awareness of this being a potentially chaotic system makes it possible to be prepared. However, another aspect of chaos means that the problems usually presented by such a system become a benefit: This is when chaos turns up in the important business of encryption.

Chaotic secrets

"In nature's infinite book of secrecy,
a little I can read."
William Shakespeare, 1564–1616

Concealing a message
Ever since information has been written down, ways have been devised to conceal it from prying eyes. Broadly, there are three mechanisms to do this: hiding the message, often in plain sight, so that only those who know how to find it can read it; using a code, where words or phrases are given a totally different meaning from their usual one; or using a cipher, where components of a message (usually, but not always, the letters and other characters) are replaced by other equivalent components according to a set of predetermined mathematical rules.

Some of the oldest methods involve the first approach—and it can still be one of the most effective, because there is no obvious message at all. An early (if slow) approach was to shave off a messenger's hair, write the message on his or her head, then wait for the hair to grow back, concealing the text until the head is shaved again. A rather quicker approach is to have a selected set of books on a shelf, placed in such an order that particular letters on the books' spines spell out a message. Without being aware that the books are being used to carry information in this way, it is practically impossible to discover the message that is being transmitted.

A	B	C	D	E	F	G	H	I	K	L	M	N	O	P	Q	R	S	T	V	X	Y	Z	W	EA	OO	SH	TH	ST	YE	CT	AV	EV	FF	ET	OV

ly	ong	ing	ence	ent	ion	ight	con	vn	and	er	itie	on	ght

Nulles.

January. Februarie. Marche. Aprile. Maye. June. Julye. August. September. October. Nouember. December. Pounds. Angels. Crownes. Ducats.

This note /. shall alwayes double the Caractere, preceding the same.

This ▫ for the puncting of periodes & Sentences / This Y for Parentheses / This ȝ to anul his precedent.

The Pope	the Earle of Arondel	the E: of Angous	Madame	waye	you
The king of France	the Earle of Oxforde	the E: of Atholl	Maiestie	receaue	your
the K. of Spane	the Earle of Sussex	the E: of Argyle	your Maiestie	day	with
the Emprour	the Earle of Northumberl	the E: of Arran	My good brother	sent	which
the K. of Demark	the E: of Leycester	the E: of Gorry	My lord	send	what
the Q. of England	the lo: H: Haward	the E: of Huntly	Maister	affect	wher
the Q. of Scotland	the E: of Shrewesbury	the b: of Glasgo	I pray you	counsell	haue
the Q. of France	the E: of Huntington	the Sp: Ambassador	most	seruice	had
the Q. mother in france	the lo: great Treasurer	the lord Seton	ernestly	support	hath
the K. of Nauarre	Sr. xpoffer Hatton	Italie	thanke	religion	shall
the Q. of Nauarre	the lord Cobham	Englande	humbly	catholik	self
the Pr: of Scotland	the Lord Hunsdon	france	humble	practise	that
the Duke of Anjou	the E: of westmerl:	Spane	comand	interprise	the
the Imperatrix	the E: of Bedford	Scotland	frende	understa	this
the Duke of Sauoye	Mr. walsingham	Ireland	matter	cause	from
the Duke of florence	the Comtesse of Shrew	flanders	Intelligence	dilligence	his
the Duke of Lorraine	the lord Talbot	the lowe cuntryes	affaire	faithfull	him
the Duke of Guyse	the lord Th: Haward	Rome	dispatche	warre	she
the Prince of Orange	Sr. wil: Cicil	London	packet	soldiors	her
the Duke of Mayne	the lord Scroope	Paris	Letter	money	any
the Duchess of Parma	the E: of Hartford	Edenburghe	answer	munitions	may
the Prince of Parma	his eldest Sonne	Barwick	bearar	armour	of
the Card: Granuelle	his second Sonne	Tynmouthe	wryte	heaven	me
the Duke of Alua	the Erle of Darby	Sheffield	secret	castel	my
the Duke of Lenox	the lord Strange	the Towre of London	aduise	shippes	to
Mons.r de Mauuissiere	Charles Arondel	the bushop of Rosse	aduertis	Crowne	by

←

Queen Mary's cipher
The cipher used by Mary, Queen of Scots, when in captivity (ca. 1586), combines simple substitutions with special symbols for frequently used words and "null symbols" which convey no information, added to confuse.

Codes have often been used by the military and by intelligence services, when there are relatively few simple concepts to be communicated. So, for example, the word HELLO might be listed in a codebook as meaning "Meet at Smith's Bar in Washington on Friday at 8:00pm." A phrase containing the word HELLO could be transmitted or put online freely: Without the codebook, the code is unbreakable. It could mean anything. However, in the absence of the codebook it can't be read by those who are supposed to understand it. Then there is always the risk that a copy of the codebook could be compromised, giving away the meaning.

Ciphers, by contrast, don't need a codebook, being merely a set of rules. The sender and receiver simply have to agree what those rules are. The simplest form is substitution by taking a fixed move through the alphabet—the so-called Caesar cipher. So, for example, if your rule is to move forward by two letters, the word HELLO becomes JGNNQ. Simple ciphers like this are very easy to break. More sophisticated ciphers make use of a key—often a word or phrase—which is "added" to the message using the number value of each letter. So, for example, if the key were COMPLEX, we would add 3 (C) to the H of HELLO, making it K, 15 (O) to E, making it T, and so on. This means that sequential letters are not enciphered using the same value—so in the HELLO example, the two Ls would not usually change to the same enciphered letter..

The best cipher is a one-time pad, where the key is only used once, and the key is a set of totally random characters. This means that the enciphered message is also random, making it impossible to break the cipher with conventional methods. The problem here is that the random key still has to be distributed to both sender and receiver of the message and can be intercepted.

A chaotic key

The current encryption used on the internet gets around this problem by using a public–private key system. This is a special kind of cipher where different keys are used to encrypt and decrypt a message. The encryption key is published to the world, so anyone wanting to send you a message can encrypt it, but only you have the decryption key, so only you can read what was written. This method is breakable, but if the private key is big enough, it is so mathematically difficult that computers would typically take centuries to decipher it without that key. By contrast, there is a growing use of encryption using the truly random values of quantum physics to generate keys in two locations simultaneously, although this is yet to be widely applied.

Getting an easy-to-use but hard-to-break encryption requires constant effort as new technology keeps making the act of breaking existing methods easier. Since around 1989 a considerable effort has been put into looking at ways to use the unpredictability of chaos to produce a form of encryption. In some cases, this has been attempted by mimicking how a chaotic system jumps around to produce pseudo-random values (which become the key) that are sufficiently unreproducible to act like a one-time pad.

The big advantage of using the chaotic function in this way over a traditional one-time pad is that instead of having to get the whole, lengthy key to both sender and receiver, all that would be needed was the details of the function, the procedure used—which of itself would be useless—and a handful of parameters that set the starting values of the function. Without those starting values it would be near-impossible to reproduce the sequence produced. Typically, the result would be a string of long numbers between 0 and 1, used as the key by taking, say, the rightmost digits of each number produced by the chaotic function in turn.

Although these numbers do appear to be very close to random, and the more decimal places used, the harder it is to crack the system, there are concerns about both the definitive security of the approach and how long it takes to crunch the numbers to encipher and decipher a message—the process tends to be quite slow, involving a lot of calculation. One other problem is that the methods developed for chaotic encryption have often been devised by mathematicians or physicists who don't have the experience to robustly check the strength of the encryption. When cryptography specialists have got their hands on the result, they can sometimes easily break through to the message.

One field where there has been particular interest in using chaos for this purpose is in image encryption. Good image encryption requires an extremely large key, and with chaotic methods, a key of any desired length can be generated from a relatively small amount of information. Images are typically encrypted either by randomly changing the values that specify the color of each pixel or by swapping pixels with random relative positions. In both cases, a number of chaotic algorithms have been championed to generate appropriate pseudo-random changes—but yet again, the cryptographic experts can often break the encryption. There is no doubt that chaos *could* provide excellent encryption; the problem seems to be in ensuring that any particular approach is sufficiently secure. The math required to check this hits up against the usual problems presented by chaos.

Algorithm
A series of logical instructions to carry out a task. Usually but not necessarily implemented in the form of a computer program.

Traffic chaos

"Traffic congestion is caused by vehicles, not by people in themselves."
Jane Jacobs, 1916–2006

It doesn't make any sense
Anyone who drives regularly will be aware of the frustrating illogic and unpredictability of traffic congestion. On a highway it's possible to be held up for long periods in a snarl-up—yet once the traffic starts moving, there appears to be no cause at the front of the congestion. And when traffic is moving slowly, in stop-start fashion in lanes, it seems almost inevitable that whichever lane you are in is the one that gets left behind by the others.

This latter effect is purely psychological, and particularly strong if there are lanes on either side of you. This is because you register the movement of other cars moving forward significantly more than your own lane, an effect that is amplified when there are at least two sets of traffic, each with a chance of moving faster than your own. It doesn't help that if one of the other lanes is blocked, they will seem to move faster than your lane as all traffic from their lane will have to move into your lane at some point.

→
Traffic congestion
A traffic jam on the freeway on Long Beach Boulevard, Commerce, California

However, the way that snarl-ups form in dense traffic is genuinely difficult to predict and analyze because the structure of the jam becomes chaotic. In many ways, the flow of traffic

down a road is similar to fluid moving through a pipe. Under normal circumstances the flow is laminar (in layers), but it can undergo a form of turbulence so that, taken as a whole system, it becomes chaotic and as a result sudden changes in speed occur, causing snarl-ups as drivers overreact, sending waves of deceleration down the line.

Butterflies on wheels

One implication of the chaotic nature that can emerge in traffic flows is a kind of butterfly effect. A single motorist may make an unexpected move—for example, a car shifting rapidly from one lane into another, causing vehicles around it to brake or swerve. The brake lights ahead will cause motorists behind to brake also, and a band of deceleration will pass rapidly back through the traffic.

This ability of drivers to respond to signals, while essential for safe driving, lends an added complexity to the normal behavior of fluid flow, making chaotic behavior more likely. This makes traffic flows resemble a type of non-Newtonian fluid which becomes much thicker or can solidify under pressure, such as custard powder or corn starch mixed with water. Like such fluids, when a close group of cars are pushed together, they tend to "thicken up," slowing the flow down the highway. For this reason, counterintuitively, more cars can sometimes be got through a particular stretch of road if they travel at a slower, steady speed, rather than driving faster but having regularly to hit the brakes heavily.

Non-Newtonian fluid
Newton made the assumption that the viscosity (the resistance to flow or gooeyness) of a fluid stayed the same under pressure. However, some fluids, such as nondrip paint and ketchup, become less viscous under pressure, while others such as custard can even become solid under pressure.

Unfortunately, the real-time nature of traffic management means there usually isn't time to run the kind of ensemble-forecasting approach used in meteorology, so the knowledge that chaotic behavior is occurring is not necessarily hugely helpful in attempting to untangle a jam; as a result, more ad-hoc approaches tend to be used. However, given significantly longer timescales to work with, the awareness of a chaotic component has proved very helpful when dealing with animal population numbers.

Population panics

"Population, when unchecked, increases in a geometric ratio. Subsistence increases only in an arithmetic ratio. A slight acquaintance with numbers will show the immensity of the first power in comparison with the second."
Thomas Malthus, 1766–1834

Fibonacci's rabbits

There is a sequence of numbers that often crops up in nature, known as the Fibonacci series. This is started off with a pair of ones, after which each subsequent number is produced by adding the previous two numbers together. So, the series runs 1, 1, 2, 3, 5, 8, 13, 21, 34, 55 .... The first-known mention of this intriguing progression was in an Indian mathematical text from before 200 BCE, but it was the thirteenth-century Italian mathematician Leonardo of Pisa, better known by his nickname Fibonacci (a contraction of "son of Bonacci" in Italian), who brought awareness of it to a Western audience.

Fibonacci wrote *Liber Abaci* (The Book of Calculation) in 1202, which discussed the series that would pick up his name. His main intent in writing this book was to bring to Europe aspects of Arab mathematics, including the use of the Indian-originated number symbols we now usually call Arabic numerals. This wasn't the first time this huge improvement on Roman numerals had been introduced to the West, but the timing was right for Fibonacci's book to become the spark to encourage the approach to be widely used. However, from our viewpoint, the

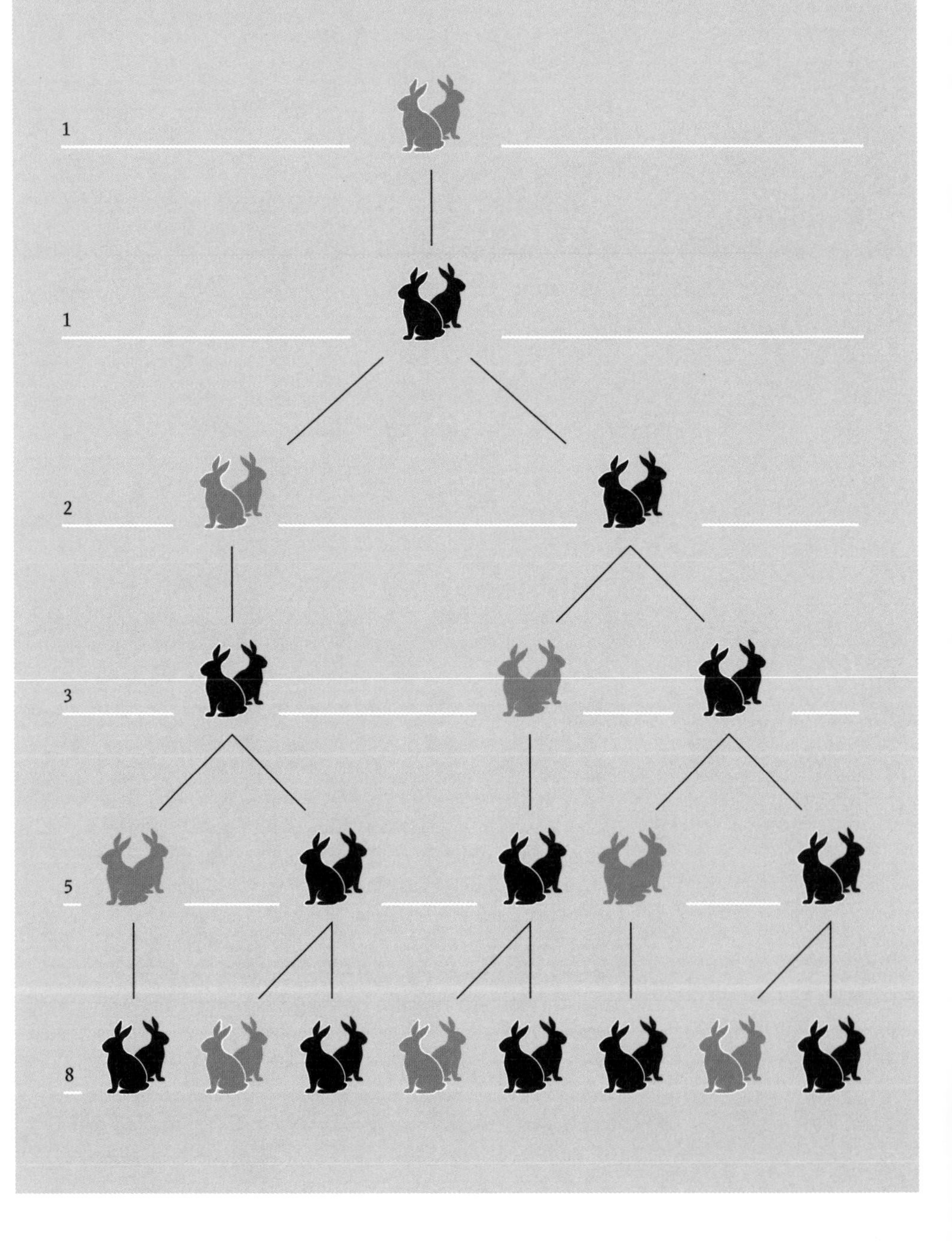
Fibonacci's rabbits
In the diagram the gray rabbits are babies and the black rabbits are mature pairs capable of breeding.
1
1
2
3
5
8

most significant content here was Fibonacci's discussion in *Liber Abaci* of a very simplistic set of breeding rabbits.

Fibonacci made use of a basic population model, starting with a single pair of baby rabbits. In his model, rabbits take one month to mature, and an adult pair of rabbits produces a pair (one male and one female) of rabbits after one month. They continue to do so after each elapsed month. No rabbits die. So, at the start of the calculation there is one pair of baby rabbits. At the end of one month, that original pair are now adult rabbits, but there are no offspring. A month later, their first babies arrive and we now have two pairs of rabbits. At the end of month three, the first pair gives birth again and the second pair matures. Now we have three pairs—two adult pairs and one baby pair. At the end of month four, the two adult pairs both produce offspring and the third pair matures. So now we have five pairs—three adult and two baby pairs. So far, the series is 1, 1, 2, 3, 5 ... and so it continues in good Fibonacci style.

Real rabbits

The Fibonacci series crops up regularly for real in nature. For example, it can be reflected in the layout of petals or seeds in a plant, due to the sequential way that the plant grows. But in practice, one place it doesn't really turn up is in animal populations. In part, this is because the populations would rapidly become uncontrollably large—the series has no end, growing ever larger and faster toward infinity. This is partly because no animals ever die in the model, but also the runaway population growth is unrealistic, because the interactions between animals, and between animals and their environment, are far more complex than the series suggests.

Even after taking these factors into account, it was assumed for a long time that population numbers would be well behaved. Aside from drastic influences—wild weather, disease outbreak, and the like—the expectation was that a population would broadly grow to the kind of level that the environment would support and then settle down to something like equilibrium. But as we have already had a hint of, in the 1970s, Australian scientist Robert May realized that something strange happens when the growth rate in a population from one measured period to the next hits a certain level.

With relatively small growth rates, things do take place as expected. The population gradually increases until it reaches what's known as the "carrying capacity"—effectively the maximum population that can comfortably be supported by a particular environment. A relatively simple equation describing this had been discovered in 1838 by Belgian mathematician Pierre-François Verhulst. But May found that if the population growth rate was set to three times over the measured period or more, something odd occurs. Rather than growing steadily until it plateaued, the number of animals would oscillate, jumping from period to period between a high value and a low one. As May gradually increased the growth rate, the values moved from this single split of values, known as bifurcation, to oscillations between four, eight, and more possibilities. It was like the period doubling of the dripping faucet (see page 116). And with a growth rate of just under 3.57 the outcome appeared to jump around totally randomly. This was classic chaotic behavior. (It was, in fact, in a paper written by American mathematicians James Yorke and Tien-Yien Li, describing May's effect, that the term "chaos" was first used in the mathematical sense.)

This tendency to move into chaos, here due to the interactions of the different factors influencing population size, has to be taken into account by scientists looking at the field of population dynamics which studies such changes in numbers of a particular organism in an environment. As usual, chaos mathematics does not allow us to make specific and accurate predictions of what is going to happen—rather it makes it clear that these are impossible—but it does mean that, for example, when a population suddenly and unexpectedly plunges in numbers, it's important not to assume that this is the result of a new and drastic external impact—it could easily be chaos in action. And probabilities can be applied to growth by modeling a large number of small differences in the starting conditions.

Modeling
An essential tool of science, modeling involves the construction of mathematical structures that behave in a similar fashion to a real-world phenomenon. Models are usually now constructed using computers and often employ simulations where random selections are made from distributions that represent the population being studied.

Population chaos is a regular factor of normal life, but the final example of chaos in action is one that we have only discovered as we push the limits of our technology: quantum chaos, and its place in electronic devices.

Quantum chaos

"It is often stated that of all the theories proposed this century, the silliest is quantum theory. In fact, some say that the only thing that quantum theory has going for it is that it is unquestionably correct."
Michio Kaku, b. 1947

1 + 1 = 3

The modern world could not function without electronics—whether it's the smartphone in your purse or the computer system that ensures the supermarket remains stocked with food. We take electronics for granted. Yet electronic devices are entirely dependent on quantum physics. Whereas some weird aspects of science and math have very little impact on day-to-day life, quantum physics, with its dependence on randomness, is a constant presence. It has been estimated that 35 percent of the GDP (gross domestic product) of developed countries has an underlying dependence on quantum physics.

As we have seen, the quantum world is in many respects the opposite of chaos. Both appear to have randomness at their heart, but where chaos is actually deterministic but impossible to predict well, quantum effects are probabilistic, but those probabilities can be known with great accuracy. Generally speaking, we expect electronic devices to always behave the same way—yet in some respects this is surprising. Quantum particles will (and do) do strange things. It's only the sheer bulk of, say, electrons in an electrical current that ensures that

Visualization of a Rydberg hydrogen atom wave function
The wave function describes the probability of finding the electron in a location. Here the electron has been boosted far above its usual level.

we don't have computers telling us that *1 + 1 = 3*. That said, in some cases quantum physics and chaos can come together to produce surprising outcomes.

Chaos on a tiny scale

Most chaotic systems we are familiar with are "macro scale"—of a size that we can see and interact with directly. However, all mechanics on the visible scale is ultimately dependent on quantum physics and there are circumstances where chaotic effects are detectable in quantum processes.

The most studied areas tend to be in the spectra produced when electrons change energy levels in atoms. When matter glows due to heat—think of a piece of metal in a forge going from dull red through orange, yellow, and finally to white—the electrons in the metal atoms are boosted up to higher energy levels by the heat, but then drop back down to a lower level. As each electron drops to a lower energy level, it gives off that energy as a photon of light—a quantum of light energy. But when some atoms with electrons boosted to unusually high levels (so-called Rydberg atoms) are in an electrical field, the production of spectra becomes chaotic, and the results unexpected.

This is because the higher the energy of an electron, the closer together are the levels it could occupy. At high energy levels they become almost a continuum, making possible the same kind of chaotic outcomes as "normal" matter. Chaotic behavior can also occur when electrons are moving through a collection of atoms—when they interact with an atom they can get temporarily caught up, and the time taken appears to be chaotic as the possible paths for the electron start to split, like the "bifurcation" experienced in some chaotic processes.

Quantum chaos only tends to crop up in relatively rare circumstances, so is unlikely to have an effect on the workings of your smartphone. But in the next chapter we are going to look to a wider picture arising from an understanding of chaos which can produce a very different outcome: complexity.

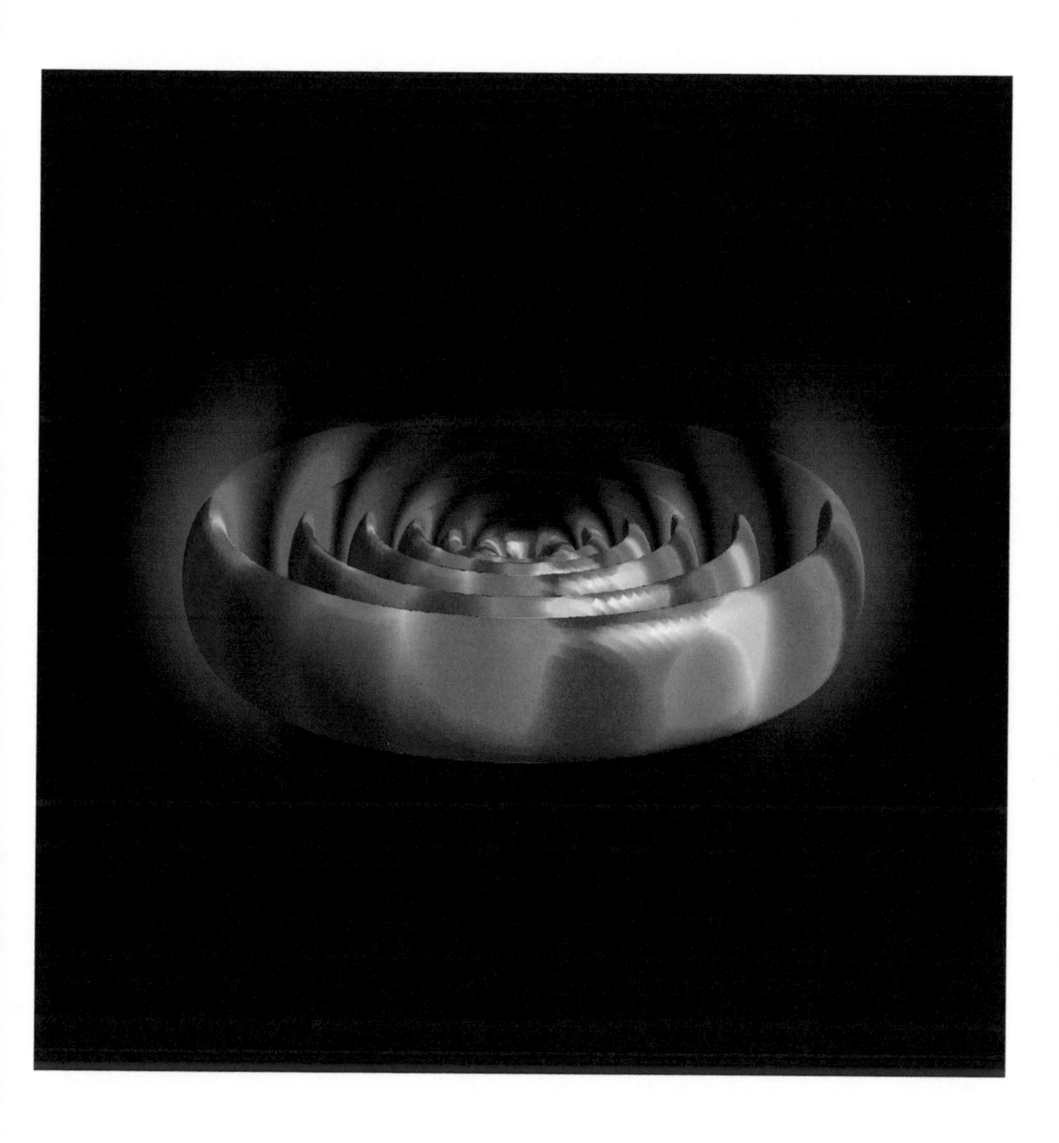

7
Complexity and Emergence

Complex systems

"Both science and art
have to do with ordered complexity."
Lancelot Whyte, 1896–1972

There's more than simple chaos
So far we have been considering chaotic systems, where the interaction between different components results in complex and unpredictable outcomes. The systems themselves need not be complex. Think, for example, of the hinged pendulum or the three-body problem. Their component parts are very simple, but the resultant behavior of the system is surprisingly complex. However, chaos is part of a bigger picture of complex systems where there is no overall control directing the outcome, but the interaction of individual parts of the system results in often surprising results.

We have already seen the way that beautiful patterns such as the Mandelbrot set images can emerge as a result of a chaotic system. Broader complexity is often typified by emergence, where new capabilities are produced by the interaction of components: with emergence, the whole is greater than the sum of the parts. Here we see systems that organize themselves without any overview or instruction. As is the case with more destructive chaotic systems, the outcome is rarely predictable because there is still a dependence on small changes in the starting conditions.

There are many mathematical and scientific ways of defining complexity. Where things change (as most systems do), the complexity of a system can be seen as the amount of detail required to describe all possible outcomes. If we think of a noncomplex system—for example a ball being thrown under a single source of gravity (or one that is so dominant that we can ignore the others, such as a ball thrown on Earth), ignoring air resistance, we can use a very simple formula to accurately predict how the ball will behave and where we will find it at a point in the future. But with a complex system, there are so many potential outcomes from interaction between parts of the system down the line that it would be very difficult to describe all the outcomes without a vast amount of information.

One important feature of many types of real-world complex systems is that they are self-organizing.

Self-organizing systems

Being self-organized typifies complex system behavior because it sounds complicated, but in reality, it can be driven by very simple factors, without any outside guidance or plan.

A simple example is what happens if you run a stream of hot water down a sloping board that is evenly covered in wax. To begin with, the water will skitter here and there over the surface in no particular pattern, but before long the wax will start to melt preferentially in some of the locations, due to small variations in the wax surface. Once channels form in the wax, they become self-reinforcing. The more hot water runs down a channel, the deeper it becomes and the better it is at carrying away water. A pattern builds itself on the surface of the wax. This pattern is caused by the uneven wax surface, but there was no plan at the outset—and it would be impossible either to predict how the channels would emerge or to generate exactly the same pattern twice. A similar effect can occur on a shoreline where rivulets in the sand self-organize.

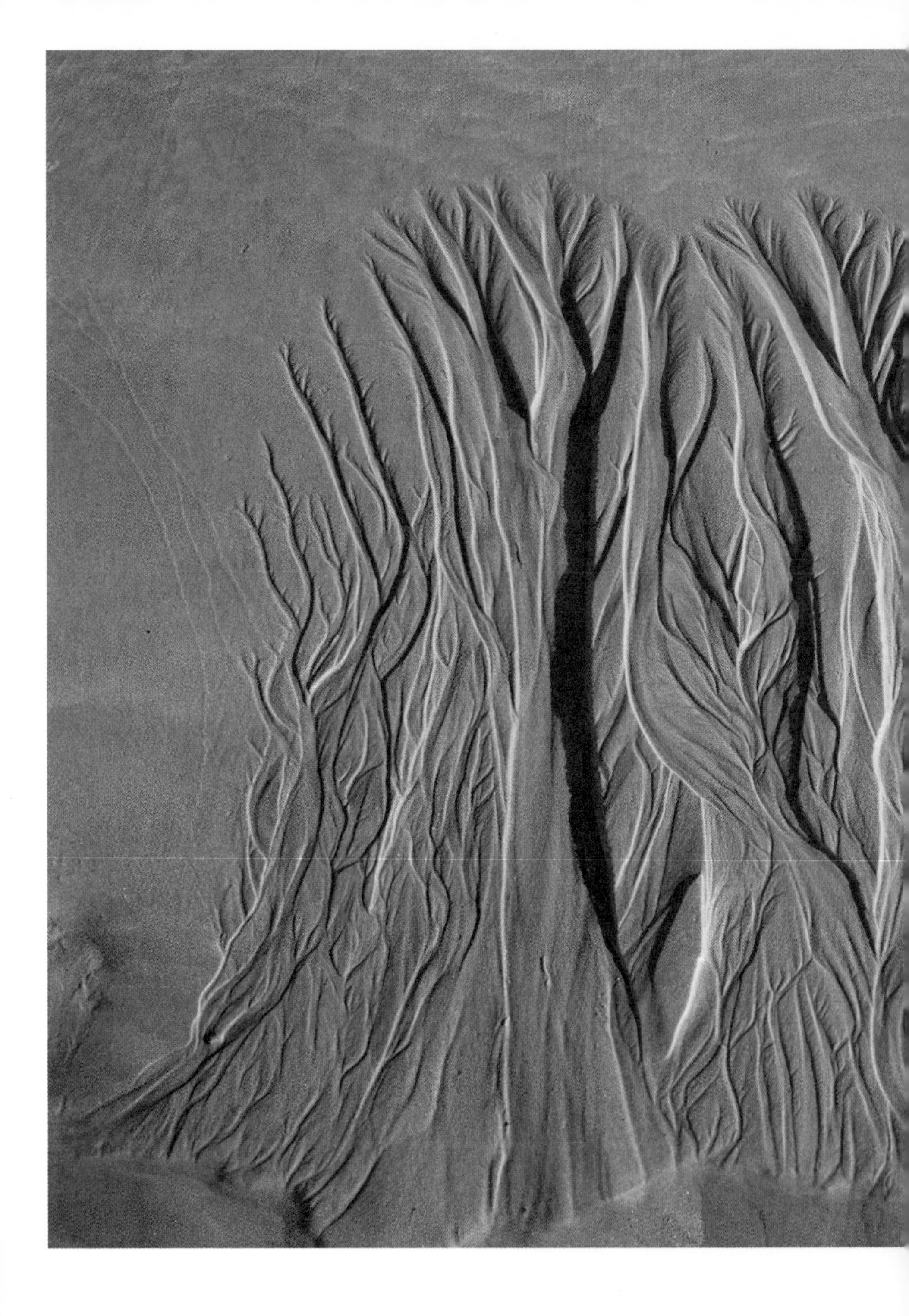

Self-patterning system
These markings left by rivulets on a beach at low tide are fractal self-organizing structures.

→

Slime molds
Despite being single-celled organisms with no nervous system or external structure, slime molds can adopt complex plantlike forms.

Another beautiful example comes from fascinating organisms known colloquially as slime mold. These are single-celled organisms that spend much of their existence floating around in an isolated fashion, with no interaction between individual cells. However, when there is not much food available, the cells start to connect with each other, forming an emergent structure that can have the appearance of a multi-celled organism. The slime mold complex can produce remarkable patterns and move around as if directed. But there is no direction behind its formation; the collective organism has no brain and no nervous system. Because the system is self-organized, though, it is capable of forming these remarkable patterns.

A similar, but far more complex thing happens within a brain (such as your own). Although there is an outline "plan" for (say) the construction of a human in the form of DNA and the way that the DNA is modified by external processes, that blueprint does not specify how the neurons in the brain are to be linked together. It's the links between neurons, known as axons, interacting with each other at junctions called synapses that are responsible for everything from your memory to your ability to think. They make you the person you are. Yet this structure isn't planned out in advance. The inner configuration of your brain is another complex, self-organizing system. In fact, the human brain is the single most complex system that is currently known.

This complexity is, in part, because of the sheer number of the connections formed in this particular self-organizing system. Your brain is thought to have at least 100 trillion synapses—possibly as many as 1,000 trillion. This reflects the power of connectivity. As you increase the number of components in an object, the number of ways to connect them together increases dramatically. So, for example, 10 objects can be connected in 45 different ways; 100 objects in 4,950; 1,000 objects in 499,500—and by the time we get to just 1 million objects, in 499,999,500,000 ways. You have around 100 billion neurons.

Just like the hot water running over the wax, the neural pathways in your brain form dynamically as time goes on—and also like the pattern in the wax, once a connection is well-established it is more likely to be used, and this use results in it becoming thicker still and easier to access. Unlike the wax, though, the process is reversible. As well as thickening when frequently engaged, underused pathways become thinner

and harder to access, particularly when an individual is under pressure. This is why it seems to be difficult to be creative under pressure: We fall back on the well-trodden pathways.

One last example of such self-organization is a localized ecology. In a particular region, different animals and plants and the physical environment form a complex system that, once again, will not normally have an outside hand organizing it, but will instead organize itself based on the interaction of the components. This is why it can be so difficult to predict the outcome of introducing a new species to a region. Think, for example, of the chaos (both literal and mathematical) caused in Australia by the introduction of the nonnative rabbit. So invasive and disruptive did the animal prove that a 330-mile (530-kilometer) rabbit-proof fence was built to keep rabbits out of Queensland, and biological warfare on the species by using the disease myxomatosis was carried out.

←

White-matter architecture
Displayed using diffusion nuclear magnetic resonance, the complex self-organizing structures support and interact with the gray matter neurons in the brain.

Chaos makes for self-organization

It might seem that such an ability to self-organize without direction is the diametric opposite of chaos—but in reality, self-organization typically emerges from chaos. In our first, simple scenario, when water is poured down a wax-covered board, the initial motion of the water is chaotic. It is almost as if the unpredictability of chaos is necessary to give self-organization a chance to emerge.

One striking example of emergence from chaos is familiar to anyone who has seen pictures of Jupiter—the Great Red Spot. At first glance, calling this reddish blob "great" seems an exaggeration—it is a relatively insignificant part of the gas giant's surface. But it is easy to forget the scale of Jupiter, which is closer to being a small, failed star than a large planet. Earth would fit easily into the Great Red Spot. But the really strange thing is that the spot has been there for hundreds of years. Why strange? Because the "surface" of Jupiter that we see is not solid. This is a *gas* giant. The apparent surface is a roiling mix of chaotic motion in the gasses that make up Jupiter.

The assumption was that Jupiter's Red Spot was a storm that had built up in those gasses, something like a planet-sized hurricane. But if this were the case, how could the Spot have possibly lasted for so long? Storms, by their nature, come and go. On Earth they rarely last for more than a few days, not (thankfully) for hundreds of years; scaling a storm up to the size of Jupiter would not give a hurricane a lifespan that extended to centuries.

It was only with the development of chaos theory that it was realized that the Great Red Spot was not a hurricane, but rather an example of the chaotic motion of the surface enabling a self-organized island of relative calm. A better model for the type of fluid flows that could produce a long-term effect like the Spot proved to be the kind of relatively stable flows that we also get in Earth's atmosphere and seas. It is not like a hurricane, but closer to the Jet Stream or a water "conveyor" system such as the Gulf Stream, that may vary in intensity and exact position from year to year but that sees consistent long-term motions of fluid in predictable directions.

The Great Red Spot is a self-organized stable region emerging from the interaction of the surrounding elements of chaos. Such self-organization is just a part of the wider phenomenon of emergence, often associated with complex systems.

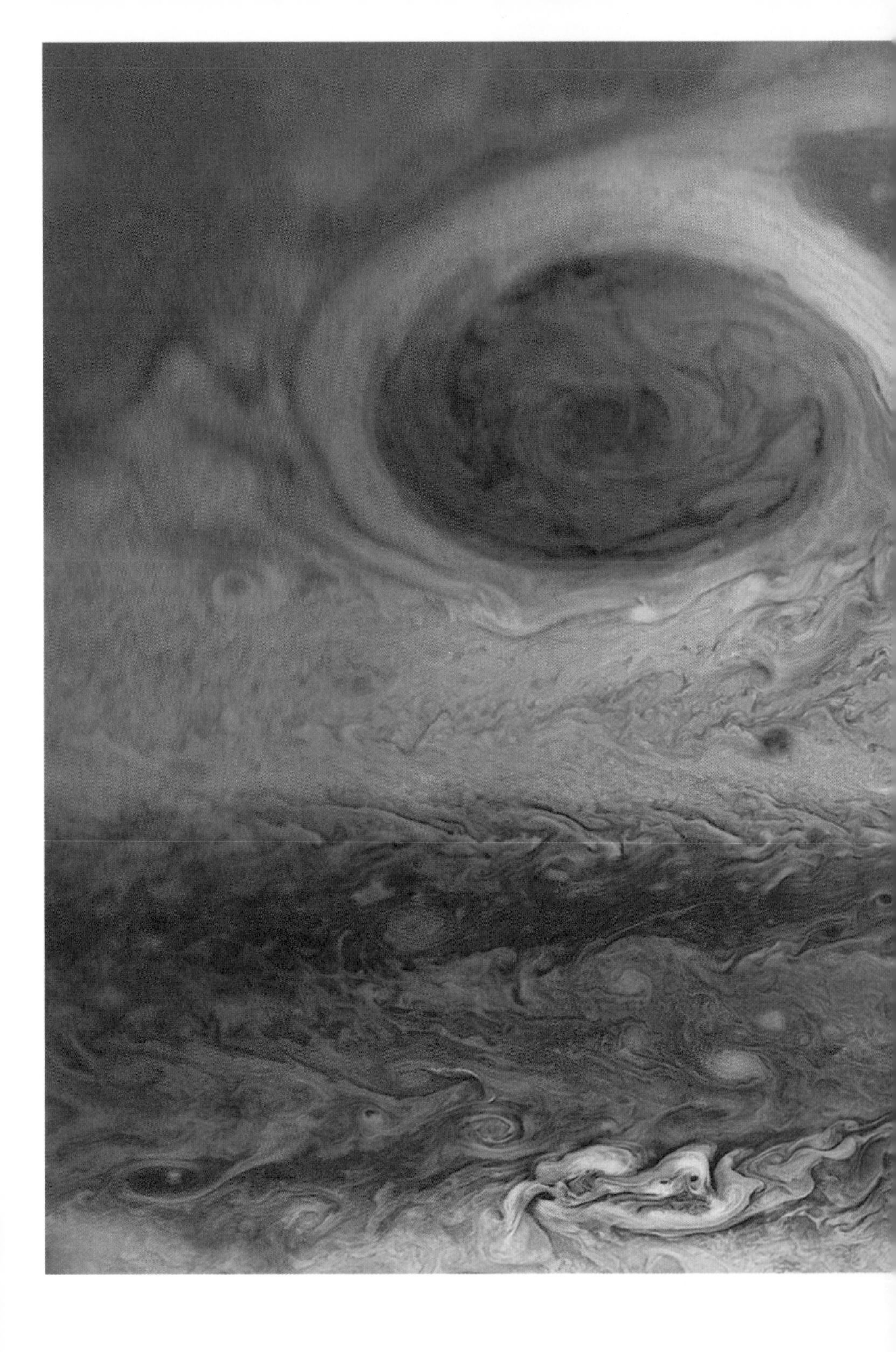

Great Red Spot
The dramatic region of self-organized stability on Jupiter that is larger than Earth.

Emergence

"A great deal of the universe does not need any explanation. Elephants, for instance. Once molecules have learnt to compete and to create other molecules in their own image, elephants, and things resembling elephants, will in due course be found roaming around the countryside."
Peter Atkins, b. 1940

Statistical emergence
Although it's not a concept that is taught at school, emergence is a common outcome of the interaction that takes place in complex systems. Perhaps the closest we come in everyday science is in statistical mechanics, used to describe the behavior of gasses.

You may remember from science lessons a number of "gas laws," which relate the interaction of three fundamental properties of a gas: its volume, its temperature, and its pressure. So, for example, Boyle's law tells us that the pressure of a gas is inversely proportional to its volume (as the volume decreases, pressure increases), while Charles's law tells us that the volume of a gas is proportional to its temperature.

However, this picture of behavior for an overall volume of gas emerges spontaneously from the random interaction of vast numbers of gas molecules, all traveling at different speeds in every possible direction, and repeatedly colliding with the other gas molecules and with the sides of the container. The regular, predictable gas laws emerge wholesale from this chaotic motion. Here, the emergence is statistical. We can calculate the outcome by looking at the distribution of the speeds of the molecules and the density of the gas. But most emergence is more sophisticated than this. Take, for example, that most extraordinary case of emergence, life itself.

More than the sum of the parts

Think about yourself for a moment. You are a living, conscious being. You are entirely made up of atoms, around 7×10^{27} of them in total. There is nothing else in your body. Those atoms are neither alive nor are they conscious. Life and consciousness are both emergent properties of the system that is you.

Emergence, as Peter Atkins's words sum up, means that the whole is more than the sum of its parts. There is more to a living organism than a collection of atoms, or of cells. This isn't a mystical observation—it's a simple physical fact. To be a human being, your atoms (or cells) need to interact with each other in an intensely complex system—and from that interaction emerges something new, which is capable of far more than a collection of individual cells or atoms ever could.

In a living organism we are dealing with extreme complexity, but even on a simple level, complex structures can emerge from interaction between adjacent components of the system that makes them up. Think, for example, of the beautiful, complex shapes of snowflakes. These filigree, six-sided structures are immensely variable, yet all emerge from the simple interaction of a small number of factors with a dash of chaos thrown in to add complexity to the mix.

A molecule of water consists of a single oxygen atom with two hydrogen atoms attached; in each molecule, the hydrogen atoms are at about a 104.5-degree angle to each other.

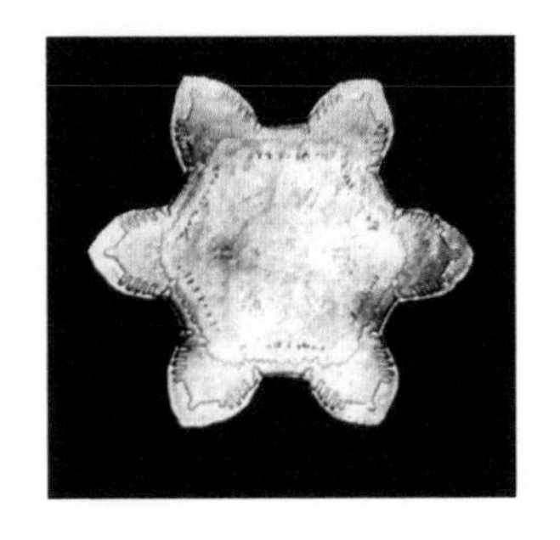

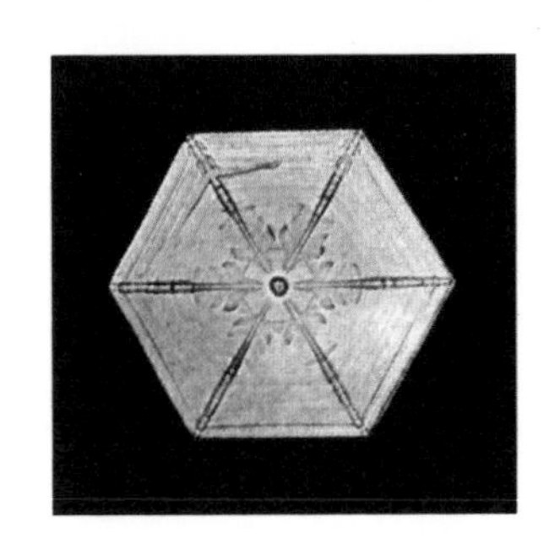

Snowflakes
Some of the many snowflake forms portrayed in Wilson Bentley's 1931 book *Snow Crystals*.

A combination of this molecular shape and the bonds that link water molecules together as they cool means that water naturally forms crystals in a six-sided lattice, and as these molecular-scale crystals grow, that six-sided form extends into the six-armed snowflake patterns that are so familiar.

The remarkable structures of snowflakes were first discovered by a Swedish cleric, Olaus Magnus, back in 1555, though the full wonder of the variety of the possible shapes was only made clear with the introduction of microscopes in the early seventeenth century. Even then, the beauty of snowflakes was not widely appreciated until 1885 when American amateur photographer Wilson Bentley started capturing images of snowflakes with early photographic technology.

Bentley produced a book of microscope photographs of snowflakes near the end of his life, in 1931. This classic work, *Snow Crystals,* contained an impressive 2,300 photographs. It was Bentley who first made the observation, based on his life work, that "no two snowflakes are alike." There is no scientific foundation for this idea, and it is easy enough to find identical flakes in some of the simpler shapes. But it is certainly true that there is a vast variety of snowflake forms.

The traditional, delicate snowflake shape with its six distinct arms (called "dendritic" meaning tree-like) grows when temperatures are particularly low; when the atmosphere is warmer, with the air not much below freezing point, the snowflakes tend to form simpler six-sided platelike crystals. The seemingly unique nature of many of the snowflake shapes is because their growth is governed by chaos. They are fractals made solid, mathematical forms with chaos at their heart.

Shoals and flocks

Just as the intricate structure of a snowflake emerges from the molecule-to-molecule interaction of water, so other emergent structures are produced by shoals of fish and flocks of birds. Each individual animal responds to other members of the flock or shoal nearby. There is no overall control, but this local-level interaction results in remarkable emergent patterns, which can shift and pulse like a living being.

The way that complex shapes can develop despite a total lack of coordination is typical of emergence. Each fish in the shoal or bird in the flock is simply reacting to the other animals in its immediate vicinity. Typically, three factors determine a shoal's or flock's behavior. Each is simple, yet collectively, by allowing these "rules" to guide their movement, members of the sinuously flowing masses of fish, or the rapidly swirling flocks of birds (delightfully known as a murmuration if they happen to be starlings), act as if they were a single, complex entity. Any individual in the group will typically only interact with a handful of near neighbors—in the case of starlings, for example, this is typically seven other birds.

The first factor influencing the formation is maintaining separation. The members of the group do not want to risk collision, and so act like electrically charged particles that repel each other if they get too close to their neighbors. Second, assuming collisions are avoided, the group members take a steer from those around them. They head in a direction that is influenced by the collective direction of their neighbors, trying broadly to follow a similar path. Finally, they head toward a midpoint between their neighbors, which enables them to handle rapid changes of direction. There are likely to be subtle variations in these rules, but it is from surprisingly simple mechanisms of this kind that the animated feel of a shoal or flock emerges.

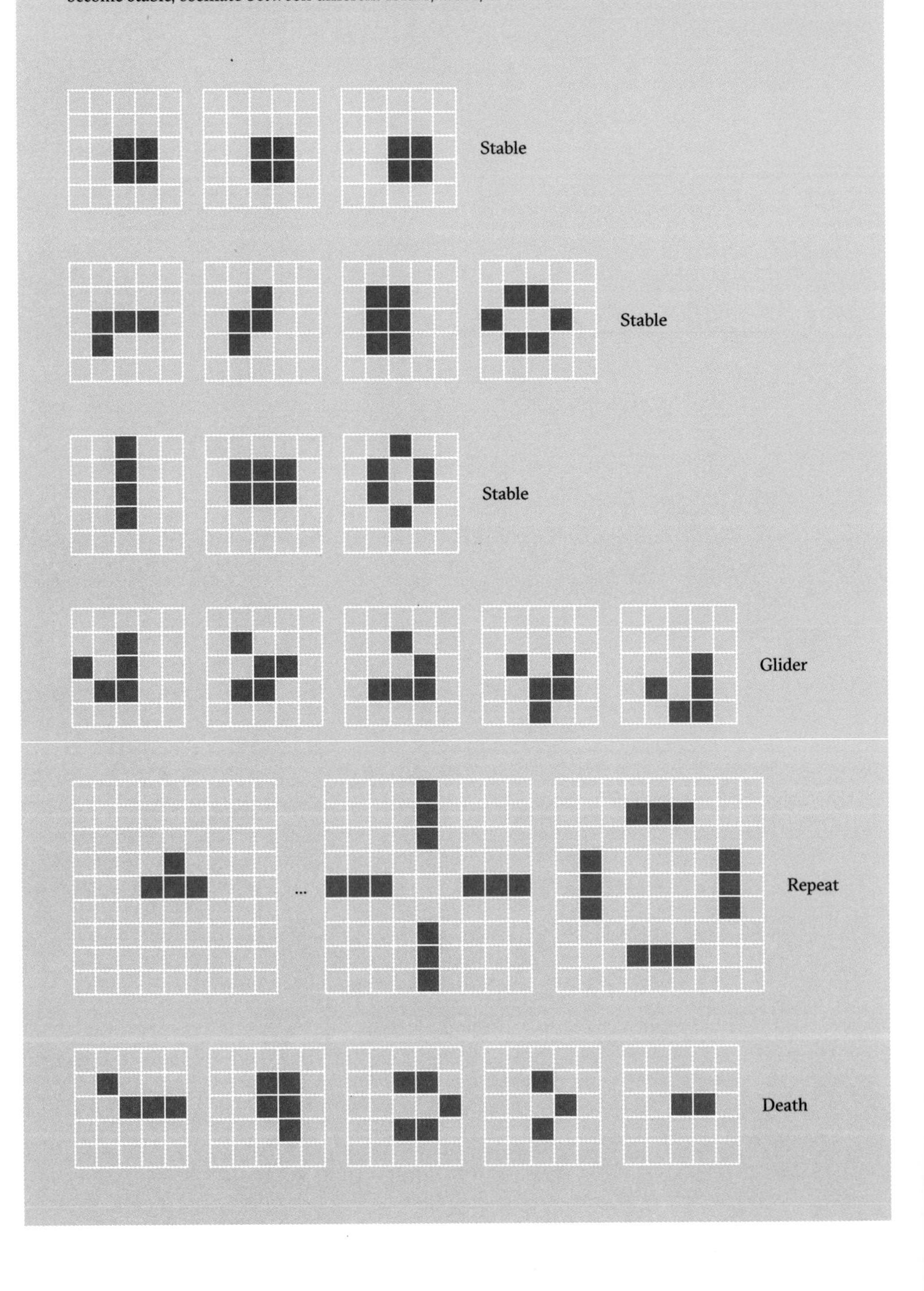
Simple Game of Life patterns
From small collections of cells in the Game of Life a range of outcomes can emerge where patterns become stable, oscillate between different forms, move, or die out.
Stable
Stable
Stable
Glider
...
Repeat
Death

Dunes and the Game of Life

An even simpler moving emergent structure than a flock comes in the form of a sand dune. There is nothing in the sand particles that specifies how the dune shape will emerge: it is the chaotic interplay of the wind that produces the initial dune. But as the shape of the dune builds, there will come a point where sand grains start to flow over from the back and down the front of the dune, shifting the whole structure along.

Even greater sophistication of progression can be seen in a computerized environment that owes something to both the mechanisms of flocks and of sand dunes, called the Game of Life. Though it is intended to reflect the real world, much of the work on chaos and complexity is based on computer models—of which Life was one of the earliest examples.

Devised by English mathematician John Conway in 1970, Life consists of a handful of simple rules that result in complex emergent behavior. An area of space is divided up into equal-sized cells. Some of these cells are initially colored in. All the game then requires is a simple set of rules to see what will emerge. Colored-in cells are considered alive and empty cells dead. The rules are that a live cell will die if it has fewer than two immediate living neighbors, or more than three. It will stay alive with two or three neighbors. And a dead cell with three living neighbors will come to life.

Given those simple rules, the state of the grid of cells is run forward in time, repeatedly applying the rules to the starting set of live cells. Some arrangements will die off entirely. Some will move around or oscillate. Others shoot out new groups of living cells, which move across the grid. Remarkable complexity of behavior emerges from extremely simple rules. Some of the structures produced by the Game of Life can look uncannily as if they have been produced intentionally—but a different level of sophistication emerges when a collection of living organisms becomes more than the sum of its parts.

Superorganisms

"After a day of raiding and destroying the edible life over a dense forest the size of a football field, the ants build their nighttime shelter—a chain-mail ball a yard across made up of the workers' linked bodies, sheltering the young larvae and mother queen at the center. When dawn arrives, the living ball melts away ant by ant as the colony members once again take their places for the day's march."
Melanie Mitchell, b. 1969

→
Living ant bridge
Brazilian army ants construct a bridge with their own bodies to cross a miniature ravine.

Joining the army
Although shoals and flocks move together with wonderful synchronicity, and the "living" cells in the Game of Life can develop surprising structures, they can't compare with the way that some species of ants, bees, wasps, and termites work together. The interaction of individuals here is so strong that they are collectively referred to as "superorganisms"—where each individual insect is arguably no more a complete organism in its own right than are the cells of your body. It's easy to accept the emergent nature of an organism when its component parts are physically joined together, but it's harder to get our heads around the idea that an individual ant, bee, wasp, or termite of a species that forms superorganisms really isn't a separate entity, but an integral part of a greater whole, enabling the colony to perform remarkable feats.

An impressive example is the Brazilian army ant. Individually or in small numbers, these insects are painfully incapable. Take a few ants and form them into a ring and they will follow each other in a circular death march until they collapse. But get them together in sufficient numbers and they become far more than individual insects. As vividly described by computer scientist Melanie Mitchell, these ants form a nest out of their own bodies. When they need to cross an ant-scale ravine—between rocks, say—rather than go the long way round, they can form a bridge with their linked bodies to allow the colony to cross the obstacle and continue with their march.

In a single, conventional organism there is usually a central brain and a nervous system that extends into the body, passing messages and controlling actions. These messages are electrochemical. For the superorganism, messages are usually passed chemically between the insects by using scent chemicals known as pheromones, as well as by specialist mechanisms such as the famous waggle dance of honey bees. In a superorganism there is no central brain—any intelligence and ability are distributed, rather as some computing systems have a distributed network, sharing the load around many small processors.

Electrochemical
In physics and chemistry we are used to electrical and chemical systems being mostly separate, but in biology many mechanisms are electrochemical, controlled or powered by the flow of electrically charged particles in fluids and across membranes.

A buzz of bees

Ants, termites, and wasps are among the best known of the superorganisms (though there are many species of each that do not operate this way). There are also some shrimp and a surprising mammal (of which more next). But the archetypal superorganism is made up of bees. We perhaps see this most obviously when part of a bee colony forms a swarm, acting briefly as what almost appears to be a single entity.

The purpose of the swarm is to enable the superorganism to reproduce. The swarm is splitting off from the original colony and establishing itself as a new one, and the formation of the swarm emphasizes the component-like nature of the individual bees. The new site will have been checked out by a group of scout bees and at the swarm's heart will be a queen bee—either the current queen, after a weight-loss program to enable the usually extremely large queen to fly, or a new queen for the nascent colony.

There is huge variety in the size and structure of colonies, but in the most familiar of bees to us, the honey bee, the colony will feature male drones, with the role of fertilizing the single queen's eggs, and a much larger number of female worker bees, responsible for collecting food, constructing the nest, and defending the colony. One of the few secondary roles the drones take part in is joining workers to help keep the nest at the right temperature, either by shivering to heat it up or by using their wings to circulate air and cool the nest down.

Bearing in mind, once more, the simplicity of an individual bee, their collective behavior (technically described as being "eusocial") makes it clear that the superorganism label is more than a metaphor. A bee colony is very much a single organism where the component parts happen to be physically separate from each other. The single active queen is more like an organ, while the individual workers act as something between an organ and a cell, cooperating with very little individual intelligence but collectively able to build complex structures, to maintain and protect the hive, and to locate and organize collection of food.

It's not that the emergence of the superorganism is more radical than, say, an individual mammal. There is far more complexity emerging from a mammal's simple collection of cells (and, for that matter of course, each individual bee also features emergent properties as a functional, if comparatively limited, organism in its own right). But it is hard not to be amazed by the achievements of a superorganism. And if bees take us by surprise, it's more remarkable still when the components of a superorganism are, like us, mammals.

The extraordinary mole rat

Most of us are at least vaguely familiar with the idea of social insects acting together to produce remarkable structures and display colony behavior, but it's not something we'd usually attribute to mammals. Obviously, there is cooperative behavior in many mammals—human beings provide the ultimate example of being able to change their environment through social cooperation. However, our ability to come together to produce results that no individual could do alone is the result of conscious communication and interaction between distinct individuals. Even when acting as a team, we are not collective parts of a superorganism with hardly any individual capabilities, as colonies of social insects are. However, there is one mammal that does form a superorganism. One look at these animals is enough to know they are unusual—but appearances can be deceiving: the naked mole rat is far more than the sum of the outward features suggested by its name.

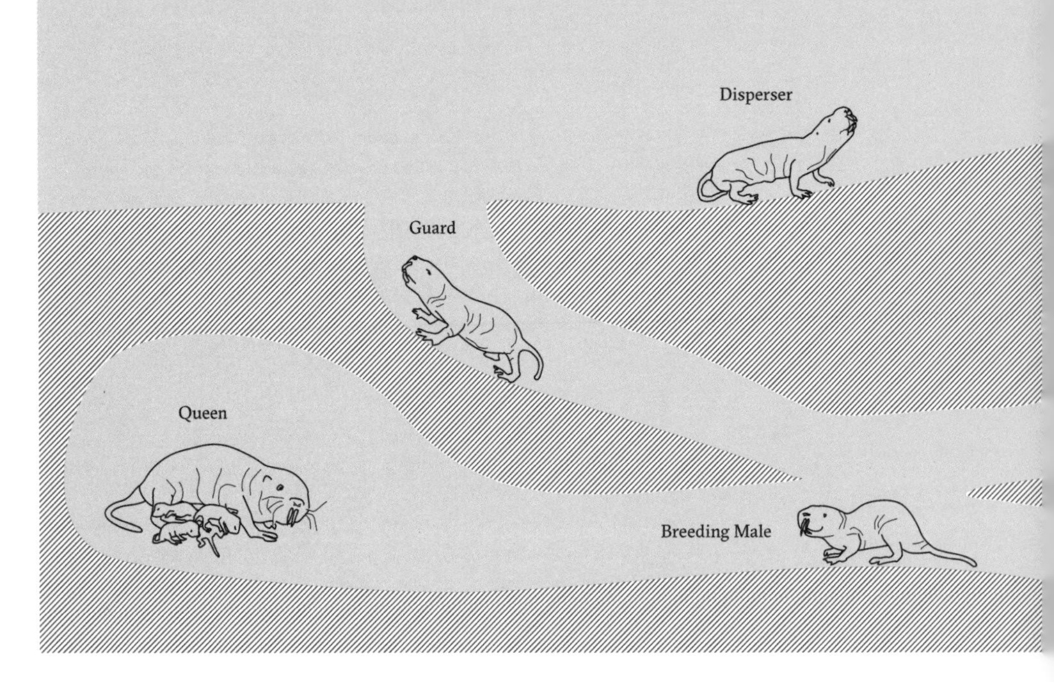

Naked mole rat colony
Most mole rats spend their entire life underground in a colony, dominated by the breeding queen.

To be honest, the naked mole rat is bizarre looking—it's like a 3D depiction of an animal that has gone horribly wrong. This is a rodent from East Africa that lives in colonies in underground burrows. The animal has lots of unusual traits—because of low oxygen levels in its tunnels it has a very low metabolism and doesn't have the normal mammalian ability to regulate its body temperature. It can't feel pain in its skin and has a remarkable ability to resist the onset of cancer. But from our viewpoint, its outstanding trait is that it forms a superorganism.

The most obvious feature of this is that a colony has a single female queen, a handful of active males, and the rest of the members are nonbreeding workers. The workers are of both sexes, but female workers do not have fully developed reproductive organs. Although the queen is not as dramatically oversized as is the case with nonmammalian queens, she is still significantly larger than the non-sexually active females.

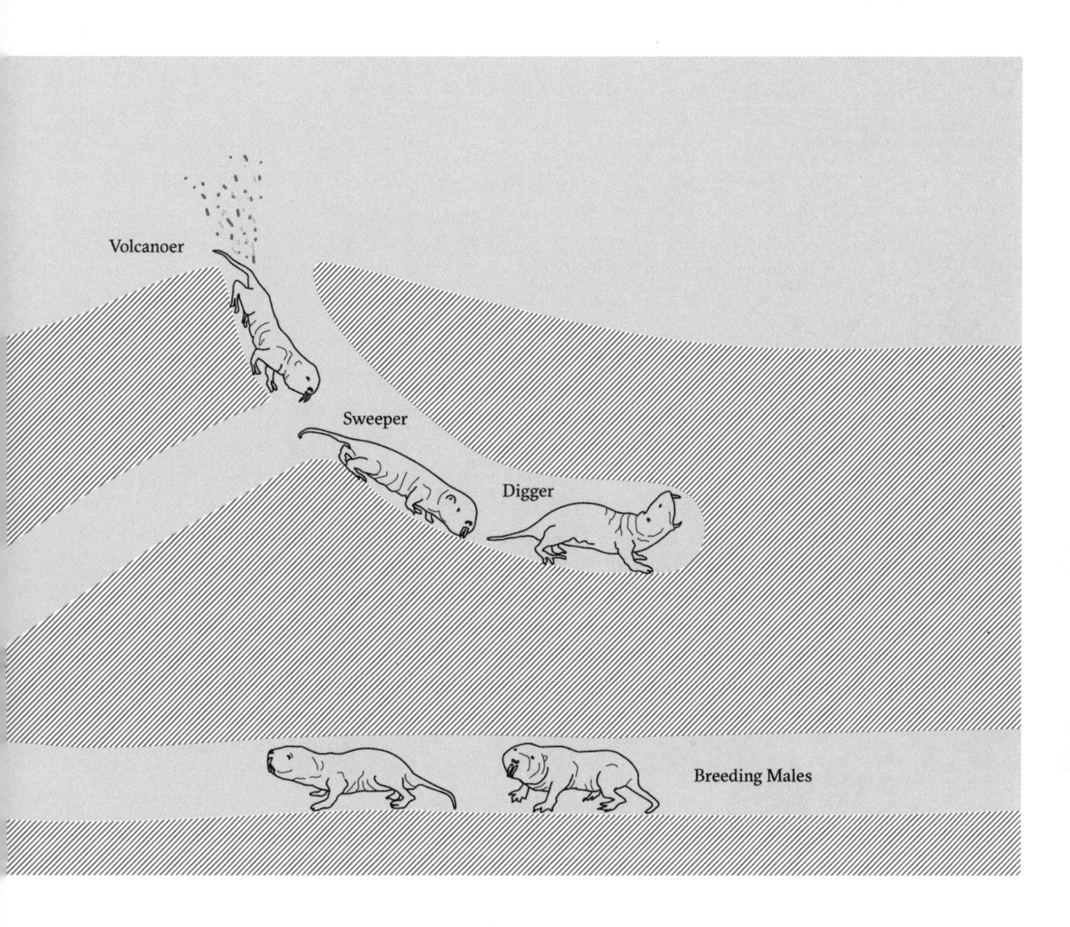

Among the workers there are a number of roles, for example those who build and maintain tunnels and those who guard the colony. There are even special versions of the male mole rat known as dispersers which are equipped with fatty supplies to enable them to journey out of the colony to reach another mole rat group in order to mate, reducing the inevitable inbreeding that usually occurs in enclosed colonies. Unlike their insect equivalent, the rest of the colony will never leave the tunnels even to collect food, as they live on tubers that they reach underground.

Apart from being a mix of fascinating and repulsive, the naked mole rat is a reminder of that essential aspect of complexity that produces the rich mix of life on our planet: adaptation.

Adaptation

"In all works on Natural History, we constantly find details of the marvellous adaptation of animals to their food, their habits and the localities in which they are found."
Alfred Russel Wallace, 1823–1913

→
Darwin's finches
Illustrations of different beak types of finches observed by Darwin on the Galápagos Islands, from his 1873 *Journal of Researches*.

In evolution's driving seat
The concept of adaptation is most strongly associated in science with evolution. One of the first clear illustrations of adaptation in this context was a driving force behind English naturalist Charles Darwin's development of the theory of evolution, arising from his observation of the birds now known as Darwin's finches.

These are a collection of species of finch found on the Galápagos islands off the coast of Ecuador. Though closely related, the different variants have distinctly diverse beak shapes. Observation of this was part of the evidence that would lead to the idea that organisms adapt to their environment from generation to generation through natural selection. Genetic variation (something Darwin was not aware of) means that some offspring would suit the environment better than others. These individuals are more likely to survive and reproduce, keeping more of the variants that are better suited to their environment. Over time, the species becomes adapted to its surroundings.

Darwin did not directly observe adaptation occurring, but over a period of 40 years, the U.S.-based English biologists Peter and Rosemary Grant repeatedly visited the island of

1. Geospiza magnirostris.
2. Geospiza fortis.
3. Geospiza parvula.
4. Certhidea olivacea.

Daphne Major in the Galápagos, recording data on the finches. Here they found clear examples of adaptation in action. For example, when the island was hit by drought in 1977, birds with bigger beaks were better able to crack the large, hard seeds that became more common under those conditions. As a result, there were more big-beaked birds over the next few years. However, a few years later, heavy rainfall made the conditions more suitable for softer, small seeds to thrive—in turn increasing the numbers of the small-billed birds, directly adapting to changing conditions.

Although evolutionary adaptation is very familiar because of its importance to biology, it is only a small subset of the larger picture of adaptive systems.

The recipe for adaptation

For adaptation to occur we need a complex system that can respond to changes, either in the environment or in parts of the system. Such responses often occur as a result of the feedback loops we have already come across, and generally require some type of memory, in the form of a medium that is able to keep information on what was or wasn't successful. (In the case of evolution, that memory is held in the specific genetic codes in the DNA that produce traits that are best suited to the environment.)

The simplest form of adaptive system is probably one we've already met—the governor, which makes changes to the system (for example by letting steam escape from a steam engine) until it reaches a state where there is no longer a prompt to action.

One simple type of adaptive system is sometimes described as being "self-adjusting." In a self-adjusting system, rather than gradually changing things as a result of feedback, the system has a memory of what has happened in the past and acts accordingly. This is particularly of interest in a system that is able to move into chaos, as self-adjusting systems can undergo a process given the rather grandiose title of "adaptation to the edge of chaos." In effect, the system moves gradually toward the border between order and chaos, but never crosses it.

Arguably the most effective adaptive system we know is the one we've already mentioned as emergent: life.

Emergent life

"The broadest and most complete definition of Life will be—The continuous adjustment of internal relations to external relations."
Herbert Spencer, 1820–1903

What is life?
Defining life is painfully difficult. On the whole it's one of those "I know it when I see it" things. We know that, say, a person, a snail, and a flower are alive, but a rock isn't. Biologists tend to avoid the question, instead listing properties of things that are alive, such as movement, nutrition, and reproduction. But one thing that is clear about life is that it is a property that emerges from some complex systems, making them adaptive.

We know that a living thing is made up of a collection of atoms. Nothing else. There is no magic extra ingredient. Yet no one would suggest that the individual atoms making up a living thing are alive. Between atoms and the whole living things are a number of structures. Molecules, for example. Again, individual molecules are not alive—even the immensely complex molecules that make up chromosomes, each a single, immensely long DNA molecule. Atoms and molecules come together to make up cells. And this time we're in a gray area.

You can argue whether cells are alive or not. They certainly exhibit some of the behaviors of life—but unless we are talking about a single-celled organism (of which there are vastly more than the complex organisms such as animals and plants), a cell

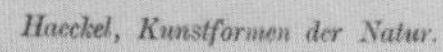

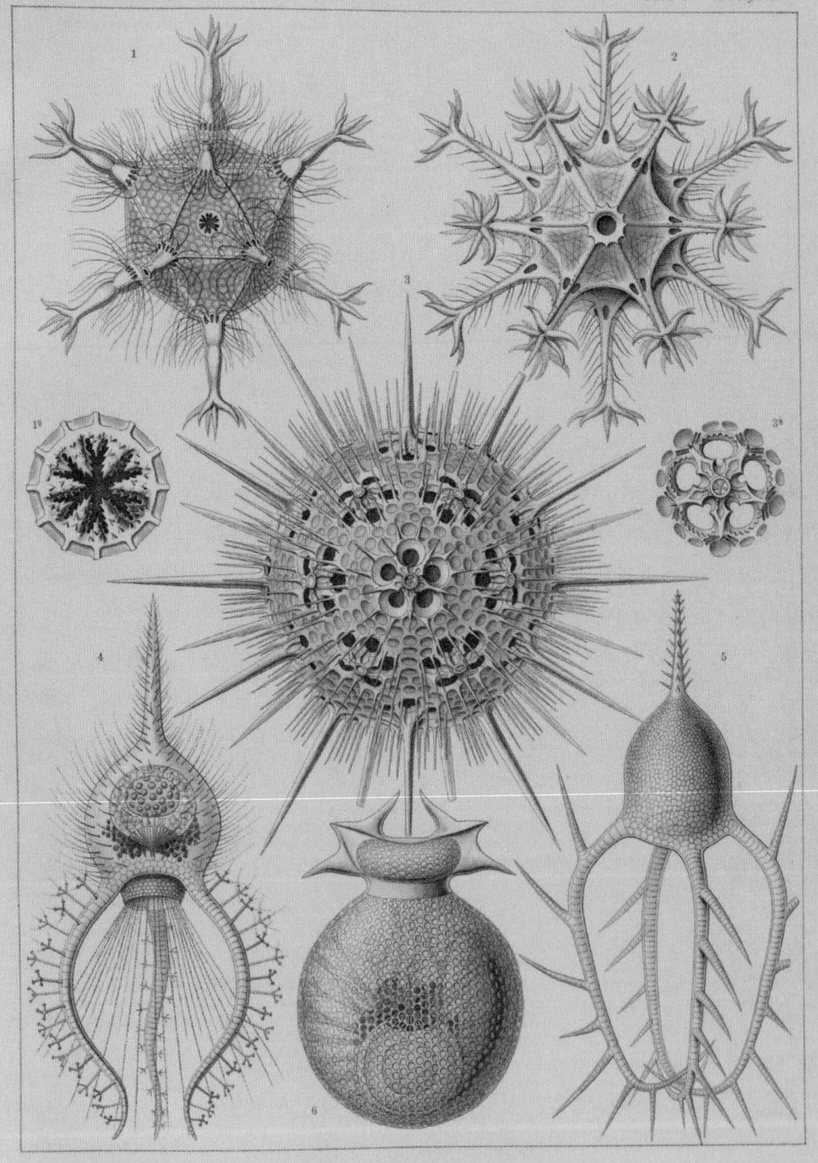

Phaeodaria. — Rohrstrahlinge.

←

Single-celled organisms
Phaeodaria from Ernst Haeckel's 1904 book *Kunstformen der Natur* (Artforms in Nature)

can't live on its own for very long. It needs other cells around it to interact with. And finally, somehow, from the reaction between the molecules in the cells, and in complex organisms between the cells, life emerges.

One way of looking at life is that there is a system that is not in equilibrium with its surroundings and stays that way: something hinted at in the words of Herbert Spencer quoted above. Normally, natural systems move into equilibrium with their surroundings. Energy passes between the system and its environment until there is a balance—think of a plate of hot food, left to stand until it reaches the same temperature as things around it. But living things stay permanently out of balance with their surroundings. They take in energy, using it to make sophisticated structures. As they increase the order within their own systems, they send out heat, causing more disorder in the world around them.

In such a system view, life is both emergent and adaptive. It is, perhaps, the definitive complex system. But some believe that we are capable of constructing something that has at least some of the complexity and emergent capabilities of a living organism.

Artificial intelligence

"Artificial intelligence is based on the assumption that the mind can be described as some kind of formal system manipulating symbols that stand for things in the world."
George Johnson, b. 1952

Adaptive robots
There is no doubt that, even without anything that could really be considered intelligence, some of our technology is adaptive. It's a fairly standard approach to robotics. For example, rather than explicitly tell a robot vacuum cleaner how to effectively clean a room, it will be typically allowed to move around randomly, sweeping as it goes. This is not efficient, but without concerns about time, it's a perfectly feasible way to achieve the task. Already there is likely to be a degree of feedback involved—for example as the cleaner approaches an obstacle it will receive feedback from its sensors to change its speed and direction. However, more is possible.

Without any plan of a room, it is easy enough for a robot to begin by moving about at random, but as it goes, it can build up a map of obstacles. As a result, it can refine its route to be more efficient and to ensure that it is capable of returning to its charging point as the energy in its batteries drops. Such a cleaner is adapting its routing to its environment. It is not being told in advance where to go, but through interaction, feedback, and memory, it adapts so it does not need to continue making the same mistakes.

A more sophisticated version of this approach is used in the now common approach to artificial intelligence we have already met—machine learning. Here, a system is not told the "rules" of its environment—it has to learn and adapt. It is allowed to do what it wants, and gets scores depending on whether an action it takes is considered good or bad. Over time, the system learns and can generate what amounts to a new emergent strategy from this process. We have already seen this applied to image recognition and discovered the potential issues encountered in making it work effectively.

In another example, machine learning systems have learned to play some games extremely well without being given the rules of the game. Again, they play over and over, initially taking random actions, but by responding to "rewards" for getting things right, they will gradually learn how to play better. Playing the classic computer game Breakout, a machine learning AI discovered a clever technique for getting massive scores by getting a ball positioned above most of the bricks that have to be blasted through in the game. This way, the ball is stuck, repeatedly knocking bricks out without requiring any action from the player.

However, it should be stressed that some have labeled this kind of machine learning "artificial unintelligence." A machine learning system *does* adapt, but it does not "know" what it is looking for. It's also worth repeating that humans can typically learn to recognize something new after seeing a handful of images—machine learning systems require thousands of images to learn in a far less reliable fashion. They don't know what they are doing, something we would tend to associate with consciousness.

Consciousness
Philosophers continue to debate whether consciousness exists at all, or whether it is an illusion that makes us think our minds are separate from our bodies. Most scientists believe that consciousness is an emergent property of the complexity of our brains.

Emergent consciousness

At present there is no suggestion that artificial intelligence systems are conscious. However, there is a rich theme in science fiction that computer networks could get increasingly complex as extra capabilities are added to them, eventually reaching a conscious state (the *Terminator* movies are a good example). Realistically, though, this is like expecting a mat of algae to become conscious as more and more cells are added—it's not enough to have connectivity of lots of basic elements. There need to be specific structures in place to provide the effect we describe as consciousness.

Despite the difficulty of producing a suitably complex system for appropriate emergence, some computer scientists do believe that it is possible to engineer a kind of consciousness. American roboticist Hod Lipson believes that the key is "self-simulation" where a robot organism has a mental model of its body's behavior, with an interaction between this and the actions of the body. He has demonstrated this principle with a robotic arm that can achieve tasks it has not been trained to carry out.

However, many believe that such devices are not achieving consciousness, they are simulating it. Apart from anything else, a conscious organism would not only be able to, for instance, pick something up without being explicitly trained to do so, but would decide that this was something that it wanted to do, rather than being told to do so. It's arguable that far more of the programming would need to be emergent to have a real shot at anything like consciousness or even basic life.

Even so, life exists. We are conscious. These properties have emerged, in part due to adaptation in the complex systems that are living things. And that's truly amazing.

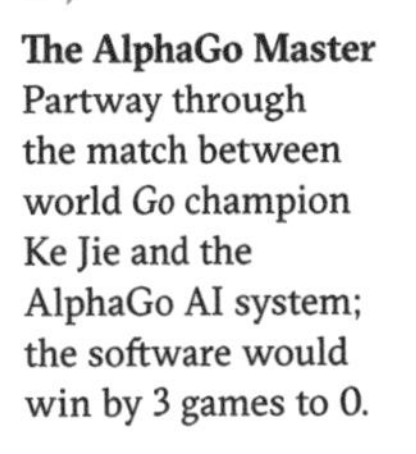

The AlphaGo Master Partway through the match between world *Go* champion Ke Jie and the AlphaGo AI system; the software would win by 3 games to 0.

Welcome to chaos and complexity

"If our brains were simple enough
for us to understand them,
we'd be so simple that we couldn't."
Ian Stewart, b. 1945

The real world is complex and chaotic
Complex systems leading to chaos and emergence are fascinating and can seem very strange. But we need to bear in mind that they are not special cases; the real world is full of such systems. In reality, complexity is the norm and chaos often ensues. It is only in the very tightly confined world that science often chooses to study—precisely because this reduces complexity—that we can put complexity to one side and ignore it.

For physicists this is often possible—though they still have to cope with turbulence, many-body problems, and more. But for biologists, economists, meteorologists, and others dealing with complex systems as a matter of course, there is no escaping complexity and chaos.

This is why, for example, it is so difficult to get good studies of dietary recommendations. In order to avoid complexity and steer clear of chaos, what scientists try to do as much as possible is remove as many factors as they can, a process known as "controlling for" whatever the factors are. So, for example, if you want to find the effect of consuming a particular substance, you get a set of subjects who are controlled for other factors—as much as possible they are all fed exactly the same way and kept in the same environment, while varying the one factor the scientists are interested in.

Unfortunately for the scientists (if fortunately for the subjects) this approach is rarely possible when the subjects are human beings. It's one thing to do this with fruit flies or mice—another with a bunch of people.

Instead, what is attempted is to modify the data to try to wipe out the impact of other factors. But often this is practically impossible. For example, we often hear that a "Mediterranean diet" is good for us. This is a diet that is high in vegetables, fruits, beans, cereals, fish, and unsaturated fats such as olive oil, but is relatively low in meat and dairy. What isn't at all clear is whether it is sufficient simply to eat the same food as those who seem to benefit from such a diet, or whether there are many other factors which are special and have an influence on their health: the climate, the culture, living near the sea, the amount of exercise they take, and so forth—these could all influence the outcome.

The marvelous Marmorkrebs

To make matters worse, we have to consider the matter of the Marmorkrebs. These are a variant of the American blue crayfish, discovered in Germany, and they are parthenogenetic. This means that they reproduce asexually: Each baby is a clone of its (female) parent. These crayfish were considered an ideal way to show how much an animal's development is influenced by genetics and how much by the environment.

In an experiment, a batch of Marmorkrebs were raised in seemingly identical circumstances. They were kept at the same temperature in the same lighting conditions with the same amount of food, and even looked after by the same person (so there was no variation in their care). As they were all clones, they should have been both genetically and environmentally identical. Yet they turned out hugely different in everything from lifespans to social interaction.

Clones
A clone begins with an identical copy of the DNA of another organism. Identical twins and the results of parthenogenetic reproduction are natural clones, while clones can also be produced artificially from an egg cell with inserted DNA.

It seems likely that the experimenters fell foul of the reality that the complex system that was the Marmorkrebs and their environment was inevitably a chaotic one. Efforts were made to keep things consistent, but very small differences in the initial conditions resulted in huge differences in the outcome. Those differences were probably both genetic (clones have subtle differences due to DNA copying errors and damage) and environmental, with tiny variations in the environment producing big differences in outcome.

No panacea—but understanding

When chaos theory, and then the mathematics of complexity, were first discovered, it was thought that this new understanding would make a huge difference to our ability to deal with a complex and chaotic world. It hasn't. Chaos is still largely unpredictable. We don't really understand how anything but the most simple emergent capabilities of systems occur. But what the theory has done is enable us to be more comfortable with the way that such systems behave—and where possible to make allowances for such outcomes, for instance by using ensemble forecasts to cope with the impact of chaos.

The theories behind chaos and complexity help us to understand why the world around us is not like our simple models of how things should behave. And, surely, that is a great thing.

Supercell storm
Chaos in action: emergent "supercell" thunderstorm structure in Kansas, USA.

Index

O

P

Q

R

S

T

V

W

Y

Z

Picture Credits

Thank you to the following sources for providing the images featured in this book:

Alamy/Steve Bloom Images: 81; BSIP SA: 220; Vasin Leenanruska: 44; Mauritius Images GmbH: 203; Mediscan: 117; Derry Robinson: 216; Scharfsinn: 169; Science History Images: 57; Marc Tielemans: 74; David Tipling Photo Library: 10; World History Archive: 166

British Library: 21

ECMWF, European Centre for Medium-Range Weather Forecasts: 111, 112, 113

Flickr/Joseph O'Brien: 219C; Katja Schulz: 219B; Bernard Spragg: 219T

Getty Images/Christophe Archambault: 46; Hill Street Studios: 152; Scott Olson: 159; Li Wenyao/ Global Times/Visual China Group: 246

Library of Congress, Washington, DC: 18, 61, 242

© Andy Lomas: 11, 49, 83, 115, 149, 183, 213

NASA: 54, 62, 104, 190, 195, 222

National Library of Austria: 23

Nature Picture Library/Floris van Breugel: 180

NOAA: 226, 227

Rijksmuseum: 7

Science Photo Library: 93

Shutterstock/Claudia G Cooper: 108; Cammie Czuchnicki: 250; Andrey Eremin: 73; Pandapaw: 16; Pjessop: 118; Sergio Sergo: 144; SIN1980: 233; Taiga: 17; Richard Wozniak: 34

US Air Force: 187

Tyler Vigen, tylervigin.com: 157

Wellcome Collection: 39, 88

Wikimedia Commons/Berndthaller: 211; Wolfgang Beyer: 139, 140, 141; EduardoMSNeves: 66; Adam Evans: 68; Ittiz: 100; Niabot: 138; Beojan Stanislaus: 137; Wikimol: 126

Every effort has been made to trace copyright holders and acknowledge the images.
The publisher welcomes further information regarding any unintentional omissions.